STUDENT'S POCKET COMPANION

JOHN D. CUTNELL
KENNETH W. JOHNSON
Southern Illinois University at Carbondale

HENRY MARK SMITH
Delgado Community College

to accompany

PHYSICS

FOURTH EDITION

JOHN D. CUTNELL
KENNETH W. JOHNSON
Southern Illinois University at Carbondale

JOHN WILEY & SONS, INC.
NEW YORK • CHICHESTER • WEINHEIM • BRISBANE • SINGAPORE • TORONTO

COVER PHOTO: *Pete Saloutos/The Stock Market*

Copyright © 1998 by John Wiley & Sons, Inc.

All rights reserved.

Reproduction or translation of any part of this work beyond that permitted by Sections 107 and 108 of the 1976 United States Copyright Act without the permission of the copyright owner is unlawful. Requests for permission or further information should be addressed to the Permissions Department, John Wiley & Sons, Inc.

ISBN 0-471-16441-0

Printed in the United States of America

10 9 8 7 6 5 4 3 2 1

Printed and bound by Courier Westford, Inc.

Preface

The *Student's Pocket Companion* is a concise summary of the material covered in *Physics* by John D. Cutnell and Kenneth W. Johnson. For each chapter of the text, the *Student's Pocket Companion* presents a corresponding chapter, beginning with a statement of learning objectives. A list of important concepts follows. Then, on a section-by-section basis, there is a summary of all concepts, including definitions, laws, theorems, and reasoning strategies. This pocket-sized book is intended for use as a convenient and quick reference guide in circumstances when the complete text is not readily available. For instance, you can use it during lectures to mark the topics covered and make brief notes in the wide margins provided for that purpose. You can also use it as a handy source book for the concepts and equations needed when working on your homework assignments, especially when you are away from your dorm room or apartment. It will also be useful when you need to plan your studying for an exam or for that last minute pre-exam review. And in the laboratory part of the course, you can refer to it for the concepts and equations needed to analyze experimental results.

Although the *Student's Pocket Companion* is a convenient reminder of what you have studied in the textbook, it is not a substitute for the textbook. There are several reasons why. First, the textbook provides detailed explanations of the topics, and the *Student's Pocket Companion* contains only brief summary sketches. Second, to help you visualize and understand the material, the textbook uses artwork and photographs, which are omitted from the *Student's Pocket Companion*. Third, to show how to apply fundamental concepts when solving problems, the textbook uses worked-out examples and presents explicit discussions of problem-solving techniques. The constraint of producing a conveniently sized volume has prevented us from including this material in the *Student's Pocket Companion*. Similarly, the textbook contains the conceptual questions and quantitative problems typically used for homework assignments, while this material is absent from the *Student's Pocket Companion*.

Additional help in solving problems can be found on the World Wide Web (http://www.wiley.com/college/cutnell), in the *Student Study Guide*, in the *Student Solutions Manual*, and in the *Learning Ware* software that is provided on the *CD-ROM* version of the textbook. The *Learning Ware* component of the *CD-ROM* is also available as a separate software package.

CONTENTS

Chapter 1	Introduction and Mathematical Concepts	1
Chapter 2	Kinematics in One Dimension	14
Chapter 3	Kinematics in Two Dimensions	23
Chapter 4	Forces and Newton's Laws of Motion	32
Chapter 5	Dynamics of Uniform Circular Motion	46
Chapter 6	Work and Energy	53
Chapter 7	Impulse and Momentum	68
Chapter 8	Rotational Kinematics	76
Chapter 9	Rotational Dynamics	87
Chapter 10	Elasticity and Simple Harmonic Motion	95
Chapter 11	Fluids	110

Chapter 12	Temperature and Heat	125
Chapter 13	The Transfer of Heat	138
Chapter 14	The Ideal Gas Law and Kinetic Theory	145
Chapter 15	Thermodynamics	154
Chapter 16	Waves and Sound	170
Chapter 17	The Principle of Linear Superposition and Interference Phenomena	183
Chapter 18	Electric Forces and Electric Fields	190
Chapter 19	Electric Potential Energy and the Electric Potential	202
Chapter 20	Electric Circuits	212
Chapter 21	Magnetic Forces and Magnetic Fields	229
Chapter 22	Electromagnetic Induction	243
Chapter 23	Alternating Current Circuits	256

Chapter 24	Electromagnetic Waves	264
Chapter 25	The Reflection of Light: Mirrors	273
Chapter 26	The Refraction of Light: Lenses and Optical Instruments	281
Chapter 27	Interference and the Wave Nature of Light	296
Chapter 28	Special Relativity	305
Chapter 29	Particles and Waves	313
Chapter 30	The Nature of the Atom	322
Chapter 31	Nuclear Physics and Radioactivity	340
Chapter 32	Ionizing Radiation, Nuclear Energy, and Elementary Particles	353

CHAPTER 1 | INTRODUCTION AND MATHEMATICAL CONCEPTS

Physics is an experimental study of natural phenomena. Physics, therefore, necessarily relies heavily on measurement. If a quantity cannot be measured, then it is meaningless to discuss it from the perspective of physics. For any physical quantity, it is important to understand both the measurement process and the units in which the quantity is measured.

In this chapter you will be introduced to Le Système International d'Unités or SI units that will be used almost exclusively throughout the course. You will review the basic notions of trigonometry that will be useful in this course, and you will find that all physical quantities can be categorized as either a scalar quantity or a vector quantity. Vector quantities have direction and must be added by special rules of addition. Vector addition is a very important skill for any student of physics and should be mastered as it will be used extensively throughout the text.

Important Concepts

- unit
- SI unit
- kilogram
- meter
- second
- dimensional analysis
- trigonometric functions
- The Pythagorean Theorem
- scalar quantity
- vector quantity
- CGS system
- base unit
- derived unit
- conversion factor
- colinear vectors
- vector components
- vector addition
- vector subtraction

1.1 THE NATURE OF PHYSICS

- Physics has grown from the human need for men and women to explain and control their physical

environment. There are only a few basic laws of physics, and they have a wide scope of application. The same laws of physics that apply to subatomic particles, atoms and molecules also apply to astronomical bodies many times larger than our sun.
- The strength of physics lies in the fact that its laws and principles are based on experimental evidence. Before any idea or concept can be accepted as a law or principle of physics, it must stand the test of experiment.
- Since physics is an experimental science, it relies heavily on the act of measurement. It is necessary, therefore, that we understand the measurement process and the units in which a measurement is made.

1.2 UNITS

- Physics uses precisely defined units of measurement. Any measurement system consists of a small number of **base units** and a larger number of **derived units.**
- Base units are defined in terms of a physical standard and the operation used to measure the quantity.

- Derived units are defined in terms of the base units.
- The text emphasizes the system of units usually referred to as **Le Système International d'Unités**, or simply **SI units**.
- The SI base units for length, mass, and time are, respectively, the meter (m), the kilogram (kg), and the second (s).
- Two other system of units sometimes used are the CGS (Centimeter-Gram-Second) units and the BE (British Engineering) units. The table below summarizes the base units in each of the three systems mentioned.

	System		
	SI	CGS	BE
Length	meter (m)	centimeter (cm)	foot (ft)
Mass	kilogram (kg)	gram (g)	slug (sl)
Time	second (s)	second (s)	second (s)

- The *meter* is defined as the distance that light travels in a time of 1/(299 792 458) second.

- The *kilogram* is defined to be the mass of a standard cylinder of platinum-iridium alloy that is kept at the International Bureau of Weights and Measures in Sèvres, France.
- The *second* is defined as the time needed for 9 192 631 770 vibrations of an atom of cesium-133 to occur.

1.3 THE ROLE OF UNITS IN PROBLEM SOLVING

- Any quantity can be expressed in any system of units. To convert a quantity from one system of units to another, it is necessary to know the appropriate conversion factor.
- A conversion factor is a fraction equal to unity. The numerator is expressed in the desired unit while the denominator is expressed in terms of the unit to be converted. The fraction is equal to unity because the numerator and the denominator are equal. For example, to convert the measurement of 4.14 m to feet, we would use the fact that 3.28 feet = 1.00 meter to construct the conversion factor

$$\left(\frac{3.28 \text{ feet}}{1.00 \text{ meter}}\right)$$

The required calculation is then

$$(4.14 \text{ meters})\left(\frac{3.28 \text{ feet}}{1.00 \text{ meter}}\right) = \boxed{13.6 \text{ feet}}$$

- Only quantities that have exactly the same units can be added or subtracted; however, quantities with different units may be multiplied or divided.
- The units on the left side of any equation must match the units on the right side. If the units do not match, either the equation is written incorrectly or the constants are not expressed in a consistent set of units.
- Each physical quantity requires a certain *type* of unit, regardless of the system of measurement. The term **dimension** is used to refer to the physical nature of a quantity and the type of unit used to specify it.
- The dimensions of the base units for length, mass, and time, are, respectively, [L], [M], and [T].
- The dimensions of derived units are combinations of [L], [M], and [T]. For example, in chapter 2, we will see that the average speed of an object as it moves between two points is the total distance that it covers divided by the time it takes to cover the

distance. The derived SI unit in which speed is measured is the meter per second (m/s), and speed, therefore, has the dimension [L/T].

1.4 TRIGONOMETRY

- **Trigonometry** is useful throughout many fields of physics. The introductory physics student should be familiar with the sine, cosine, and tangent functions of the angle θ. These quantities can be defined in terms of the angle θ in the right triangle.

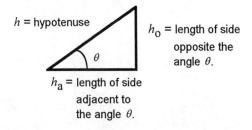

- Using the right triangle shown above,

$$\sin \theta = \frac{h_o}{h} \qquad (1.1)$$

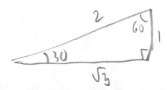

$$\cos\theta = \frac{h_a}{h} \qquad (1.2)$$

$$\tan\theta = \frac{h_o}{h_a} \qquad (1.3)$$

- Once the lengths of any two sides of a right triangle are known, the angle itself can be obtained using inverse trigonometric functions. That is,

$$\theta = \sin^{-1}\left(\frac{h_o}{h}\right) \qquad (1.4)$$

$$\theta = \cos^{-1}\left(\frac{h_a}{h}\right) \qquad (1.5)$$

$$\theta = \tan^{-1}\left(\frac{h_o}{h_a}\right) \qquad (1.6)$$

- Another useful relationship between the sides and hypotenuse of the right triangle is **Pythagorean's Theorem**:

$$h^2 = h_o^2 + h_a^2 \qquad (1.7)$$

- When the lengths of two sides of a right triangle are know, Pythagorean's theorem can be used to find the length of the unknown side.

1.5 THE NATURE OF PHYSICAL QUANTITIES: SCALARS AND VECTORS

- Every physical quantity encountered throughout this course can be categorized as either a scalar quantity or a vector quantity.
- A **scalar quantity** is a physical quantity that has only a characteristic size or **magnitude.** A scalar quantity is completely specified by giving its magnitude and the appropriate unit. Some examples of scalar quantities include mass, time, speed, electric charge and temperature.
- A **vector quantity** is a physical quantity that has both magnitude and direction. To describe a vector quantity both its magnitude and direction must be specified with the appropriate unit. Several examples of vector quantities include force, velocity, acceleration, electric fields and magnetic fields.
- Vector quantities are often represented by arrows, the length of the arrow being proportional to the magnitude of the vector and the direction of the arrow indicating the direction of the vector.

- As an algebraic symbol, vector quantities are denoted in **bold face** characters (e.g., **A**) to distinguish them from scalar quantities which are denoted in *italics* (e.g., *s*). In a hand-written algebraic equation, vectors are usually denoted by placing an arrow over the variable (e.g., $\vec{A}$).

1.6 VECTOR ADDITION AND SUBTRACTION

- Two or more vectors may be added by arranging the vectors in a tail-to-head fashion. The **vector sum** or **resultant** is obtained by connecting the tail of the first vector to the head of the last vector.
- In general, the magnitude of the resultant vector is *not* the sum of the magnitudes of the vectors being added.
- In the special case where the vectors are along the same line or *colinear*, the magnitude of the resultant *is equal to* the sum of the magnitudes of the vectors. With the tail of the second vector located at the head of the first vector, the resultant is exactly equal to the length of the two vectors. As shown below, **A** + **B** = **R** and the

magnitude of **R** is equal to the sum of the magnitudes of **A** and **B**.

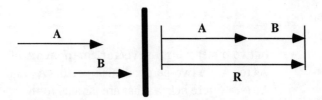

- Multiplying a vector by a scalar factor of –1 reverses the direction of the vector. The subtraction of a vector is defined as the addition of a vector that has been multiplied by –1. In the figure below, **C** = **A** – **B**. Note that this operation is the addition of the vectors **A** and –**B**; that is, **C** = **A** + (–**B**).

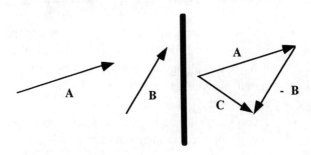

1.7 THE COMPONENTS OF A VECTOR

- In two dimensions, the **vector components** of a vector **A** are two perpendicular vectors $\mathbf{A}_x$ and $\mathbf{A}_y$ (see figure below) that are parallel to the x and y axes, respectively, and add together vectorially so that $\mathbf{A} = \mathbf{A}_x + \mathbf{A}_y$.

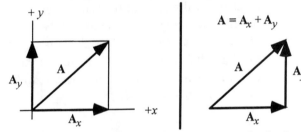

- The **scalar component** A_x has a magnitude that is equal to that of $\mathbf{A}_x$ and is given a positive sign if $\mathbf{A}_x$ points along the $+x$ axis and a negative sign if $\mathbf{A}_x$ points along the $-x$ axis. The scalar component A_y is similarly defined.
- Two vectors are equal in two dimensions if, and only if, the x components of each vector are equal and the y components of each vector are equal.
- A vector is zero if, and only if, each of its vector components is zero.

1.8 ADDITION OF VECTORS BY MEANS OF COMPONENTS

- The components of a vector provide the most efficient method for adding (or subtracting) any number of vectors.
- Suppose that **A** and **B** are two vectors to be added; that is, $\mathbf{R} = \mathbf{A} + \mathbf{B}$. Then,
  $R_x = A_x + B_x$ and $R_y = A_y + B_y$.

  The magnitude and direction of the resultant **R** can be found from Pythagorean's theorem and the inverse tangent relationship:

$$R = \sqrt{R_x^2 + R_y^2} \quad \text{and} \quad \theta = \tan^{-1}\left(\frac{R_y}{R_x}\right)$$

CHAPTER 2 | *KINEMATICS IN ONE DIMENSION*

Mechanics is the branch of physics that deals with the study of motion. Mechanics can be subdivided into two branches: *kinematics*, the study of the description of motion, and *dynamics*, the study of the causes of motion.

In this chapter, you will begin the study of kinematics in one dimension (that is, for motion along a straight line). You will be introduced to the concepts of displacement, velocity and acceleration. The special case of motion with constant acceleration will be developed in detail, and the resulting equations will be applied to the situation of a freely falling body.

Important Concepts
- mechanics
- kinematics
- displacement
- average speed
- average velocity
- average acceleration
- instantaneous speed
- instantaneous velocity
- instantaneous acceleration

- free-fall motion

2.1 DISPLACEMENT

- **Displacement** is a vector that points from an object's initial position toward its final position.
- Displacement is a vector quantity. Its magnitude is equal to the shortest distance between the two positions. Its direction is given by an arrow that points from the initial position to the final position.
- The SI unit for displacement is the meter (m). It is a length, and its associated dimension is [L].

2.2 SPEED AND VELOCITY

- The **average speed** of an object is the distance traveled by the object divided by the time required to cover the distance.

$$\text{Average speed} = \frac{\text{Distance covered}}{\text{Elapsed time}} \quad (2.1)$$

- Average speed is a scalar quantity. It has no associated direction.
- The SI unit for average speed is the meter per second (m/s). Average speed is a length divided by time; therefore, its dimension is [L/T].
- Speed is a useful concept because it indicates how fast an object is moving; however, it does *not* reveal anything about the direction of motion.
- The **average velocity** $\bar{\mathbf{v}}$ of an object is defined as the object's displacement $\Delta \mathbf{x}$ divided by the elapsed time Δt.

$$\bar{\mathbf{v}} = \frac{\mathbf{x} - \mathbf{x}_0}{t - t_0} = \frac{\Delta \mathbf{x}}{\Delta t} \quad (2.2)$$

- Average velocity is a vector quantity. It has the same direction as the direction of the displacement vector.
- The SI unit for average velocity is the meter per second (m/s). Its dimension is [L/T].
- The **instantaneous velocity**, or the velocity at a given instant, can be found from the average velocity in the limit that Δt becomes infinitesimally small:

$$\mathbf{v} = \lim_{\Delta t \to 0} \frac{\Delta \mathbf{x}}{\Delta t} \qquad (2.3)$$

- The SI unit and dimension for instantaneous velocity are the same as those for average velocity.

2.3 ACCELERATION

- The **average acceleration** is defined as the change in velocity over a particular time interval divided by the elapsed time:

$$\bar{\mathbf{a}} = \frac{\mathbf{v} - \mathbf{v}_0}{t - t_0} = \frac{\Delta \mathbf{v}}{\Delta t} \qquad (2.4)$$

- Average acceleration is a vector that points in the same direction as the *change* in velocity.
- The SI unit for average acceleration is the meter per second per second (m/s^2). Its dimension is [L/T^2].
- The velocity of an object may change (i.e., an object may accelerate) because the velocity changes in magnitude, in direction, or in both magnitude and direction. In chapter 5 you will encounter a special kind of motion called

uniform circular motion in which an object travels at constant speed in a circular path. Such an object is accelerating even though its speed is constant because the velocity vector is changing direction.

- When Δt becomes infinitesimally small, the average acceleration becomes equal to the **instantaneous acceleration**:

$$\mathbf{a} = \lim_{\Delta t \to 0} \frac{\Delta \mathbf{v}}{\Delta t} \qquad (2.5)$$

- The SI unit and dimension for instantaneous acceleration are the same as those for average acceleration.

2.4 EQUATIONS OF KINEMATICS FOR CONSTANT ACCELERATION

- When an object begins at $x_o = 0$ at time $t = 0$, and moves with constant acceleration along a straight line, the instantaneous velocity v, the initial velocity v_o, the acceleration a, and the elapsed time t are related by the following **equations of kinematics**:

$$v = v_0 + at \qquad (2.4)$$

$$x = \frac{1}{2}(v_o + v)t \qquad (2.7)$$

$$x = v_0 t + \frac{1}{2}at^2 \qquad (2.8)$$

$$v^2 = v_0^2 + 2ax \qquad (2.9)$$

- The equations of kinematics have the same algebraic form for either horizontal or vertical motion provided that the acceleration remains constant during the motion.
- In one dimension, motion in one direction is arbitrarily chosen to be positive. Any motion in the opposite direction is, therefore, negative. The directions of the displacement vector, the velocity vector and the acceleration vector are then specified with plus (+) and minus (−) signs.

2.6 FREELY FALLING BODIES

- In **free-fall motion**, an object experiences

negligible air resistance and a constant acceleration due to gravity.

- All objects in free fall at the same location above the earth have the same acceleration due to gravity.
- The **acceleration due to gravity** is a vector that points toward the center of the earth and is denoted by the symbol g. Near the surface of the earth, g has a magnitude that is approximately constant and is given by

$$g = 9.80 \text{ m}/\text{s}^2 \quad \text{or} \quad 32.2 \text{ ft}/\text{s}^2$$

- The equation of kinematics (with constant acceleration) may be applied to free-fall motion since g is nearly constant. It is natural to use the symbol y for the displacement that occurs, since the motion occurs in the vertical or y direction.
- The motion of an object that is thrown upward and eventually returns to earth contains two important symmetries. A time symmetry exists in that the time required for the object to reach its maximum height equals the time for it to return to its starting point. A type of symmetry involving the speed also exists. The speed of the object at any given value of y, while the object is rising, is equal to the speed of the object at the same value of y as the

object is falling. These symmetries arise because the object loses 9.80 m/s in speed each second on the way up and gains back the same amount each second on the way down.

2.7 GRAPHICAL ANALYSIS OF VELOCITY AND ACCELERATION

- Graphical techniques are useful in understanding the concepts of velocity and acceleration.
- It follows from the definition of average velocity that the **slope of the line tangent to the displacement versus time curve** at a particular value of the time t is equal to the instantaneous velocity at that time.

$$\text{Slope of tangent line} = \frac{\Delta x}{\Delta t} = v$$

- For example, suppose an object moves according to the x versus t curve shown at the right. The value of the instantaneous velocity of the object at time t_1 is equal to the slope of the line drawn tangent to the curve at t_1.

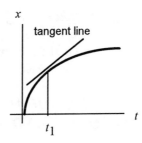

- Similarly, the **slope of the line tangent to the velocity versus time curve** at a particular value of the time t is equal to the instantaneous acceleration at that value of t.

Slope of tangent line $= \dfrac{\Delta v}{\Delta t} = a$

CHAPTER 3 | *KINEMATICS IN TWO DIMENSIONS*

In this chapter, the concepts of displacement, velocity, and acceleration, along with the equations of kinematics, are extended to motion in two dimensions. Two applications are discussed in detail: projectile motion in two dimensions and relative motion in both one and two dimensions.

Important Concepts

- displacement vector
- velocity vector
- acceleration vector
- projectile motion
- relative velocity

3.1 DISPLACEMENT, VELOCITY AND ACCELERATION

- Motion in two dimensions is described by the same physical quantities that were defined in Chapter 2 to describe one dimensional motion.
- The **displacement vector** is used to describe the change in position of an object. Suppose an object moves in the x-y plane. Let $\mathbf{r_o}$ represent the position of an object at time t_o when the object is at the point **A**. Suppose that the object then travels along the curved path shown below for a time t. Let $\mathbf{r}$ represent its position vector at the later time t when the object has reached the point **B**.

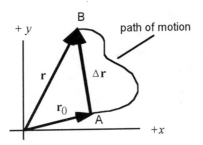

The displacement of the object is then calculated from

Displacement =

$$\Delta \mathbf{r} = \mathbf{r} - \mathbf{r}_o$$

- The **average velocity** of the object is defined by

$$\bar{\mathbf{v}} = \frac{\mathbf{r} - \mathbf{r}_0}{t - t_0} = \frac{\Delta \mathbf{r}}{\Delta t} \qquad (3.1)$$

- The velocity of the object at a particular instant is the object's **instantaneous velocity** and is defined by

$$\mathbf{v} = \lim_{\Delta t \to 0} \frac{\Delta \mathbf{r}}{\Delta t}$$

- At any instant, the velocity vector **v** has a vector component $\mathbf{v}_x$ along the x direction and a vector component $\mathbf{v}_y$ along the y direction.
- If the velocity of the object changes with time, the object has an acceleration. The **average acceleration** is defined in the same manner as it is for one dimensional motion:

$$\bar{\mathbf{a}} = \frac{\mathbf{v} - \mathbf{v}_0}{t - t_0} = \frac{\Delta \mathbf{v}}{\Delta t} \qquad (3.2)$$

- The acceleration of the object at a particular instant is the object's **instantaneous acceleration** and is defined by

$$\mathbf{a} = \lim_{\Delta t \to 0} \frac{\Delta \mathbf{v}}{\Delta t}$$

- At any instant, the acceleration vector **a** has a vector component $\mathbf{a}_x$ along the x direction and a vector component $\mathbf{a}_y$ along the y direction.

3.2 EQUATIONS OF KINEMATICS IN TWO DIMENSIONS

- Motion in two dimensions can be analyzed by treating the x and y components separately. That is, the x part of the motion occurs exactly as it would if the y part did not occur at all. Similarly, the y part of the motion occurs exactly as it would if the x part of the motion did not exist.
- A problem dealing with two-dimensional motion can be considered as two one-dimensional problems (see, for example, Example 1 in the text).

- The **equations of kinematics** for motion with constant acceleration in two dimension are given in the table below for an object that begins at $\mathbf{r}_o = 0$ at time $t_o = 0$.

x Component

$$v_x = v_{0x} + a_x t \qquad (3.3a)$$

$$x = \tfrac{1}{2}(v_{0x} + v_x)t \qquad (3.4a)$$

$$x = v_{0x} t + \tfrac{1}{2} a_x t^2 \qquad (3.5a)$$

$$v_x^2 = v_{0x}^2 + 2 a_x x \qquad (3.6a)$$

y Component

$$v_y = v_{0y} + a_y t \qquad (3.3b)$$

$$y = \tfrac{1}{2}(v_{0y} + v_y)t \qquad (3.4b)$$

$$y = v_{0y} t + \tfrac{1}{2} a_y t^2 \qquad (3.5b)$$

$$v_y^2 = v_{0y}^2 + 2 a_y y \qquad (3.6b)$$

- The directions of the components x, y, v_{0x}, v_{0y}, v_x, v_y, a_x, and a_y are specified by assigning a plus (+) or minus (−) sign to each one.

3.3 PROJECTILE MOTION

- **Projectile motion** is a kind of two-dimensional motion that occurs when the moving object (the projectile) experiences only the acceleration due to gravity, which acts in the vertical direction.
- The acceleration of the projectile has no horizontal component ($a_x = 0$). Thus projectile motion is characterized by motion with constant velocity in the horizontal or x direction and motion with constant acceleration in the vertical or y direction.
- Given the initial position and velocity of the projectile, the kinematic equations can be used to determine the position and velocity of the projectile at any later time t.
- The path followed by an object executing projectile motion is a parabola. The figure below shows two possible parabolic paths. In **A**, the object is given a horizontal initial velocity ($v_{o_y} = 0$). In **B**, the

object is given an initial velocity at an angle θ. That is, in **B**, the initial velocity has both horizontal and vertical components.

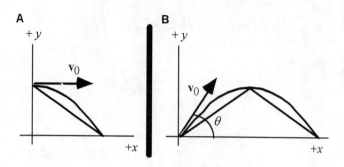

- The horizontal component of the projectile's velocity remains constant at all times and is equal to its initial value ($v_x = v_{o_x}$), while the vertical component changes because of the acceleration due to gravity.
- The speed of a projectile at any location along its trajectory is the magnitude v of its velocity vector at that location. It is given by
$v = \sqrt{v_x^2 + v_y^2}$.
- When the projectile in **B** reaches its maximum height, the vertical component of its velocity is momentarily zero ($v_y = 0$); however, the horizontal component of its velocity remains equal to its initial value ($v_x = v_{o_x}$).
- The speed of the projectile in **B** at a given height above the ground is the same on the

upward and downward parts of the trajectory. The velocities are different, however, since the velocity vectors point in different directions.

*3.4 RELATIVE VELOCITY

- When an object in motion is observed by an observer who is also in motion, the motion of the object will generally appear to be quite different than it does to a ground-based observer.
- Suppose that an object **O** is moving relative to a ground based observer **G**. The object is observed by a second observer **M** who is also moving relative to the ground-based observer. The velocity of the object **O** relative to the moving observer **M** can be calculated by using the following relationship:

$$\mathbf{v}_{OM} = \mathbf{v}_{OG} + \mathbf{v}_{GM} \qquad (3.7)$$

where
$\mathbf{v}_{OM}$ = velocity of **O** relative to **M**
$\mathbf{v}_{OG}$ = velocity of **O** relative to **G**
$\mathbf{v}_{GM}$ = velocity of **G** relative to **M**.
- Equation 3.7 is a vector equation and holds for

motion in both one and two dimensions.
- In general, the velocity of object **O** relative to object **S** is always the negative of the velocity of object **S** relative to **O**:

$$\mathbf{v}_{OS} = -\mathbf{v}_{SO}.$$

CHAPTER 4 | FORCES AND NEWTON'S LAWS OF MOTION

Chapter 4 begins the study of dynamics — the branch of mechanics that studies and analyzes the causes of motion. You will encounter the concept of *force* and use Newton's laws to determine how forces influence the motion of objects.

Important Concepts

- force
- Newton's first law
- contact forces
- non-contact forces
- mass
- inertia
- inertial reference frame
- Newton's second law
- external forces
- internal forces
- free-body diagrams
- Newton's third law
- gravitational force
- Newton's law of universal gravitation

- strong nuclear force
- electroweak force
- weight
- normal force
- apparent weight
- static and kinetic friction forces
- coefficient of friction
- tension force
- equilibrium
- nonequilibrium

4.1 THE CONCEPTS OF FORCE AND MASS

- A **force** is a push or a pull that one object exerts on another object. It is a vector quantity with both a magnitude and direction.
- **Contact forces** arise from the physical contact between two objects. **Non-contact forces** arise in situations in which two objects exert forces on one another, even though they are not touching.

4.2 NEWTON'S FIRST LAW OF MOTION

- Often, several forces act simultaneously on a body. The **net force** that acts on the body is the vector sum or resultant of all the individual forces that act on the body.
- **Newton's first law of motion** or **law of inertia** states that an object continues in a state of rest (zero velocity) or in a state of motion at a *constant speed along a straight line* unless compelled to change that state by a net force.
- Note that a net force is *not* required to sustain the velocity of an object. Rather, a net force is required to *change* the velocity of an object.
- **Inertia** is the natural tendency of an object to remain at rest or in motion at a constant speed along a straight line. The **mass** of an object is a quantitative measure of its inertia.
- Mass is a scalar quantity. The SI unit for mass is the kilogram (kg). It is a base unit with the dimension [M]. The CGS and BE units for mass are the gram (g) and the slug (sl), respectively.
- An inertial reference frame is one in which Newton's law of inertia is valid.

4.3 NEWTON'S SECOND LAW OF MOTION

- **Newton's second law of motion** states that when a net force $\Sigma \mathbf{F}$ acts on an object of mass m, the acceleration $\mathbf{a}$ of the object is related to the net force by:

$$\Sigma \mathbf{F} = m\mathbf{a} \qquad (4.1)$$

- The SI unit of force is the newton (N). It is a derived unit; in terms of SI base units, $1 \text{ N} = 1 \text{ kg} \cdot \text{m/s}^2$. Force has the dimension of $[M][L][T]^{-2}$.
- The BE unit of force is defined to be the pound (lb).
  From Newton's second law, it follows that $1 \text{ lb} = 1 \text{ sl} \cdot \text{ft/s}^2$.
- The net force $\Sigma \mathbf{F}$ in Equation 4.1 includes only the **external forces** or the forces that the environment exerts on the object.
- **Internal forces** are forces that one part of the object exerts on another part of the object. They are not included in Equation 4.1.
- A **free-body diagram** is a diagram that

represents the object and all the external forces acting on it. A free-body diagram is useful in determining the net force that acts on an object.

4.4 THE VECTOR NATURE OF NEWTON'S SECOND LAW OF MOTION

- The net force $\Sigma \mathbf{F}$ in Newton's second law has components ΣF_x and ΣF_y, while the acceleration **a** has components a_x and a_y. Therefore, in two dimensions, Newton's second law can be written in terms of two equations, one for the x components and one for the y components.

$$\Sigma F_x = ma_x \qquad (4.2a)$$

$$\Sigma F_y = ma_y \qquad (4.2b)$$

- The components will be positive or negative numbers, depending on whether they point along the positive or negative x or y axis.

4.5 NEWTON'S THIRD LAW OF MOTION

- **Newton's third law of motion**, often called the **action-reaction law**, states that when one object exerts a force (the action) on a second object the second object exerts an oppositely directed force (the reaction) of equal magnitude on the first object.
- Note that Newton's third law implies that a *single isolated force* can *never* exist anywhere in the universe. All real forces are members of an action-reaction pair.

4.6 TYPES OF FORCES: AN OVERVIEW

- There are two general types of forces that occur in nature: fundamental and non-fundamental.
- **Fundamental forces** are forces that are truly unique in that all other forces can be explained in terms of them. The three fundamental forces are the gravitational force, the strong nuclear force, and the electroweak force.

- The **strong nuclear force** plays a primary role in the stability of the atomic nucleus.
- The **electroweak force** is a single force that manifests itself in two ways. One manifestation is the *electromagnetic force* that electrically charged particles exert on one another; the other manifestation is the *weak nuclear force* that is operative in the radioactive decay of certain nuclei.
- Except for the gravitational force, all forces discussed in Chapter 4 are non-fundamental; they are manifestations of the electromagnetic force.

4.7 THE GRAVITATIONAL FORCE

- **Newton's law of universal gravitation** states that every particle in the universe exerts an attractive force on every other particle. For two particles that are separated by a distance r and have masses m_1 and m_2, Newton's law of universal gravitation states that the magnitude of the attractive force is

$$F = G \frac{m_1 m_2}{r^2} \qquad (4.3)$$

while its direction lies along the line that joins the particles.

- The constant G is called **the universal gravitational constant** and has the value 6.67×10^{-11} N·m² / kg².

- The **weight** of an object on earth is the gravitational force that the earth exerts on the object. Equation 4.3 can be used to express the weight W of an object on earth:

$$W = G \frac{M_E m}{r^2} \qquad (4.4)$$

where M_E is the mass of the earth and m is the mass of the object.

- The weight of an object can also be expressed in terms of Newton's second law incorporating the acceleration g due to gravity

$$W = mg \qquad (4.5)$$

- A comparison of Equations 4.4 and 4.5 shows that

$$g = G \frac{M_E}{r^2}$$

4.8 THE NORMAL FORCE

- A surface exerts a force on an object with which it is in contact. The component of this force that is perpendicular to the surface is called the **normal force**. The component of this force that is parallel to the surface is called the **frictional force.**
- The **apparent weight** of an object is the force that an object exerts on the platform of a scale and may be larger or smaller than the true weight if the object and the scale are accelerating.
- The apparent weight and true weight are related according to

$$\underbrace{F_N}_{\text{Apparent weight}} = \underbrace{mg}_{\text{True weight}} + ma \qquad (4.6)$$

4.9 STATIC AND KINETIC FRICTIONAL FORCES

- The **force of static friction** between two surfaces opposes any impending relative motion between the two surfaces.
- The magnitude of the static frictional force depends on the magnitude of the applied force and can assume any value up to a maximum value of

$$f_s^{MAX} = \mu_s F_N \qquad (4.7)$$

where μ_s is the **coefficient of static friction** and F_N is the magnitude of the normal force.
- Note that Equation 4.7 relates only the magnitudes of f_s^{MAX} and F_N. *It implies nothing about the directions of the vectors.*
- The **force of kinetic friction** between two surfaces sliding against one another opposes the relative motion of the surfaces. The magnitude f_k of the kinetic frictional force is given by

$$f_k = \mu_k F_N \qquad (4.8)$$

where μ_k is the **coefficient of kinetic friction** and F_N is the magnitude of the normal force.
- The coefficient of kinetic friction, like the coefficient of static friction, depends only on the nature and composition of the two surfaces. Under most circumstances the values of the coefficients of friction are independent of the apparent macroscopic area of contact between the objects and independent of the relative motion of the objects.
- For the same two surfaces, the coefficient of kinetic friction is usually smaller than the coefficient of static friction. This implies that a larger force is required to start an object moving on a rough surface than is required to keep it moving.

4.10 THE TENSION FORCE

- The term **tension** is commonly used to describe the tendency of a rope or wire to be pulled apart due to forces that are applied at each end.
- Through the force of tension, a rope can transmit a force from one end to the other.
- When a rope is accelerating, the tension force is transmitted undiminished only if the rope is massless.

4.11 EQUILIBRIUM APPLICATIONS OF NEWTON'S LAWS OF MOTION

- An object is in **equilibrium** when the object moves with zero acceleration. Thus, an object in equilibrium may be moving at a constant velocity or at rest with zero velocity.
- From Newton's second law, it follows that the net external force that acts on an object in equilibrium is zero. That is, the forces acting on an object in equilibrium must balance.
- Thus, in two dimensions, the equilibrium condition is expressed by the equations

$$\Sigma F_x = 0 \qquad (4.9a)$$

$$\Sigma F_x = 0 \qquad (4.9b)$$

4.12 NONEQUILIBRIUM APPLICATIONS OF NEWTON'S LAWS OF MOTION

- When an object is accelerating, it is not in equilibrium. Newton's second law must be used to account for the accelerations, and the motion is governed by Equations 4.2a and 4.2b:

$$\Sigma F_x = ma_x \quad (4.2a) \quad \text{and} \quad \Sigma F_y = ma_y \quad (4.2b)$$

- *Reasoning Strategy*
  Applying Newton's Laws
  Step 1: Select the object or objects (the "system") under consideration.
  Step 2: Draw a free-body diagram for each object selected in Step 1. *Include only forces that act on the object (do not include forces that the object exerts on its environment).*
  Step 3: Choose a convenient set of x and y axes for each object and resolve all forces in the free-body diagram into components that point along these axes.
  Step 4: For nonequilibrium situations, apply Equations 4.2a and 4.2b to the

object in question. For equilibrium situations, apply Equations 4.9a and 4.9b.

Step 5: Solve the two equations obtained in Step 4 for the desired, unknown quantities.

CHAPTER 5 | *DYNAMICS OF UNIFORM CIRCULAR MOTION*

This chapter introduces the kinematics and dynamics of an object traveling in a circular path. Four applications are discussed: motion of a vehicle around a banked curve, the motion of orbiting satellites, apparent weightlessness, and vertical circular motion.

Important Concepts

- uniform circular motion
- period
- centripetal acceleration
- centripetal force
- Apparent weightlessness
- vertical circular motion

5.1 UNIFORM CIRCULAR MOTION

- **Uniform circular motion** is the motion of an object traveling at a constant (uniform) speed on a circular path.
- The **period** T of the motion is the time required to make one revolution.
- The constant speed v, the period T, and the radius r of the circular path are related according to

$$v = \frac{2\pi r}{T} \qquad (5.1)$$

- In uniform circular motion, the velocity vector (which is always parallel to the instantaneous direction of motion) is always changing direction; hence, the object is accelerating.

5.2 CENTRIPETAL ACCELERATION

- The acceleration of an object in uniform circular motion is called **centripetal acceleration.** The adjective *centripetal* refers to the direction of the acceleration and literally means "center-seeking."
- The magnitude of the centripetal acceleration of an object moving with constant speed v on a circular path of radius r is given by

$$a_c = \frac{v^2}{r} \qquad (5.2)$$

- The centripetal acceleration vector always points *toward the center of the circle* and continually changes direction as the object moves.

5.3 CENTRIPETAL FORCE

- From Newton's second law, it follows that whenever an object accelerates, there must be a net force to create the acceleration.

- The **centripetal force** is the name given to the net force that gives rise to a centripetal acceleration, and it follows from Newton's second law that its magnitude is given by

$$F_c = \frac{mv^2}{r} \qquad (5.3)$$

- Like the centripetal acceleration, the centripetal force always points toward the center of the circle and continually changes direction as the object moves.
- Note that the term "centripetal force" does not refer to a new force of nature. All of the forces encountered in Chapter 4 (gravitational, tension, friction, normal forces) can be centripetal forces.

5.4 BANKED CURVES

- One example of uniform circular motion involves the motion of a vehicle as it rounds a banked curve.
- The angle θ at which a friction-free curve is banked depends on the radius r of the curve and the speed v with which the curve is to be

negotiated, according to

$$\tan\theta = \frac{v^2}{rg} \qquad (5.4)$$

5.5 *SATELLITES IN CIRCULAR ORBITS*

- Another example of uniform circular motion is the motion of a satellite in a circular orbit about the earth.
- *There is only one speed that a satellite can have at a given altitude if the satellite is to remain in a circular orbit.*
- The speed v and period T of a satellite in a circular orbit about the earth depend on the radius r of the orbit according to

$$v = \sqrt{\frac{GM_E}{r}} \qquad (5.5)$$

and

$$T = \frac{2\pi r^{3/2}}{\sqrt{GM_E}} \qquad (5.6)$$

where G is the universal gravitational constant and M_E is the mass of the earth.

Note that Equation 5.5 is independent of the mass of the satellite. Thus, for a given orbit, *a satellite with a large mass has exactly the same orbital speed as a satellite with a small mass.*

5.6 APPARENT WEIGHTLESSNESS AND ARTIFICIAL GRAVITY

- As discussed in Section 4.8, the apparent weight of an object is the force that an object exerts on the platform of a scale. In a freely falling elevator, the apparent weight of an object is zero. This is because both the scale and the object fall together with the same acceleration, and, therefore, cannot push against one another.
- Objects in uniform circular motion continually accelerate or "fall" toward the center of the circle in order to remain on the circular path. Thus, objects in an orbiting satellite "fall" with the same acceleration toward the center of the orbit and cannot push against one another. The apparent weight in the satellite is zero, just as it is in a freely falling elevator.

5.7 VERTICAL CIRCULAR MOTION

- When the speed of travel of an object on a circular path changes from moment to moment, the motion is said to be *nonuniform*.
- One example of nonuniform circular motion is the vertical circular motion of a cyclist performing the "loop-the-loop" trick. As the cycle goes around the vertical circular path, the magnitude of the normal force changes because the speed changes and the weight does not have the same effect at each point on the circle.

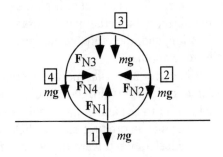

CHAPTER 6 | *WORK AND ENERGY*

Energy is one of the most useful concepts in physics. Chapter 6 introduces the concept of work done by a constant force and the related concept of energy. The strategy of the text is to introduce the two forms of mechanical energy: kinetic and potential. With these two forms, it is possible to discuss the transfer of energy from one form to another. When only conservative forces act on a system, and no work is done on the system by external nonconservative forces, the total mechanical energy of the system is conserved. The principle of conservation of mechanical energy can be used to solve a variety of problems in a much simpler manner than by using Newton's laws directly.

Important Concepts

- work done by a constant force
- kinetic energy
- work-energy theorem
- work done by the force of gravity
- gravitational potential energy

- conservative forces
- nonconservative forces
- total mechanical energy
- the principle of conservation of mechanical energy
- average power
- other forms of energy and the conservation of energy
- work done by a variable force

6.1 WORK DONE BY A CONSTANT FORCE

- When a constant force **F** acts on an object as the object moves through a displacement **s**, the force does **work**

$$W = (F\cos\theta)s \qquad (6.1)$$

  where θ is the angle between the force and displacement vectors.
- Work is done only by the component of the force **F** in the direction of the displacement. The force component that is perpendicular to the displacement does no work.

- The SI unit of work is the joule (J).
  In terms of SI base units, $1 \text{ J} = 1 \text{ kg} \cdot \text{m}^2/\text{s}^2$.
- Work is a scalar quantity; it has no associated direction.
- Work can be positive or negative. Work is positive when the force has a component in the same direction as the direction of the displacement. In this case, the angle θ between the force **F** and the displacement **s** is less than 90°. The work is negative when the force component points in the direction opposite to the displacement; that is when θ is greater than 90°. When $\theta = 90°$, the force is perpendicular to the displacement and does zero work.

6.2 THE WORK-ENERGY THEOREM AND KINETIC ENERGY

- The kinetic energy of an object with mass m and speed v is defined by

$$\text{KE} = \tfrac{1}{2}mv^2 \qquad (6.2)$$

- Kinetic energy is energy that is associated with the motion of the object.
- Kinetic energy depends on the mass m of the

object, which is always positive, and on v^2, which is always positive; therefore, the kinetic energy of an object is always positive (or zero when $v = 0$).

- Kinetic energy is a scalar; it has no associated direction.
- The SI unit for kinetic energy is the joule (J). Work and kinetic energy have the same SI unit.
- The **work-energy theorem** states that the work W done by the *net external force* acting on an object equals the difference between the object's final kinetic energy KE_f and initial kinetic energy KE_0.

$$W = KE_f - KE_0 = \tfrac{1}{2}mv_f^2 - \tfrac{1}{2}mv_0^2 \qquad (6.3)$$

- A moving object has kinetic energy because work was done to accelerate the object from rest to a speed v. Conversely, an object with kinetic energy can perform work by pushing or pulling on another object.
- If the net external force does positive work, the kinetic energy of the object increases; if the net external force does negative work, the kinetic energy decreases.
- Strictly speaking, the work-energy theorem applies only to a single particle or to an object that can be treated as a single particle. If the object becomes distorted due to the application

6.3 GRAVITATIONAL POTENTIAL ENERGY

- The **work done by the force of gravity** on an object of mass m is

$$W_{\text{gravity}} = mg(h_0 - h_f) \qquad (6.4)$$

where h_0 and h_f are the initial and final heights of the object, respectively.
- Only the difference in the vertical distances, $(h_0 - h_f)$, need be considered when calculating the work done by gravity.
- It is assumed that the difference in heights is small compared to the radius of the earth, so that the magnitude g of the acceleration due to gravity does not vary over the vertical distance $(h_0 - h_f)$. For positions close to the earth's surface, $g = 9.8$ m/s^2.
- The distances h_0 and h_f need not be measured from the surface of the earth. They can be measured relative to any conveniently chosen zero level.

- An object may possess energy by virtue of its position or configuration relative to other objects or particles with which it interacts. Such energy is called *potential energy*. Potential energy may be thought of as energy that is stored in the system of interacting objects or particles. Different types of potential energy may be distinguished by the nature of the forces involved in the interactions. For example, the potential energy associated with the gravitational interaction between a planet and a nearby object is called the *gravitational potential energy*. In Chapter 10 you will be introduced to the *elastic potential energy* of a system composed of an object attached to a spring, while in Chapter 19 you will encounter the *electric potential energy* that arises from the interaction of two or more electrically charged particles.
- **Gravitational potential energy**, PE, is the energy that an object of mass m has by virtue of its position relative to the surface of the earth. For an object near the surface of the earth, its gravitational potential energy is given by

$$\text{PE} = mgh \qquad (6.5)$$

where h is the height of the object relative to an arbitrary zero level.
- Like work and kinetic energy, the SI unit of

gravitational potential energy is the joule (J).
- Gravitational potential energy is a scalar quantity.
- Gravitational potential energy depends on both the object and the earth (m and g, respectively), and the height h. Thus, gravitational potential energy belongs to the object and the earth as a system (although one frequently refers to the object as possessing the potential energy).

6.4 CONSERVATIVE FORCES AND NONCONSERVATIVE FORCES

- The text provides two versions for the definition of a *conservative force*.
  Version 1: A force is a **conservative force** when the work it does on a moving object is independent of the path of the motion between the object's initial and final positions.
  Version 2: A force is a **conservative force** when it does no net work on an object moving around a closed path, starting and finishing at the same point.
  The two versions of the definition are equivalent. If a force satisfies the requirements of *Version 1*, then it will automatically satisfy

the requirements for *Version 2*. Conversely, if a force does not satisfy the requirements stated in *Version 1*, it will not satisfy the requirements stated in *Version 2*.

- With each conservative force, we may associate a potential energy. For example, the force of gravity is a conservative force, and the associated potential energy is the gravitational potential energy. Two other conservative forces that will be encountered later in the text are the elastic force of a spring and the electrical force of electrically charged particles.

- Many forces encountered in everyday situations are not conservative forces. A force is **nonconservative** if the work it does on an object moving between two points depends on the path of the motion between the points.

- Two common nonconservative forces are the force of kinetic friction and air resistance. In most instances, the forces of kinetic friction and air resistance point opposite to the direction of motion of the object, and therefore, do negative work on the object. Between any two points, more work is done over longer paths between the points; thus, the work done depends on the specific path traveled by the object.

- The work-energy theorem can be expressed in a form that involves the work W_{nc} done by all the external nonconservative forces. In terms of the kinetic and potential energies:

$$W_{nc} = \underbrace{(KE_f - KE_0)}_{\text{Change in kinetic energy}} + \underbrace{(PE_f - PE_0)}_{\substack{\text{Change in} \\ \text{gravitational} \\ \text{potential energy}}} \quad (6.7a)$$

or

$$W_{nc} = \Delta KE + \Delta PE \quad (6.7b)$$

where ΔKE and ΔPE are the changes in the kinetic energy and potential energy of the object, respectively.

6.5 THE CONSERVATION OF MECHANICAL ENERGY

- The **total mechanical energy** E is the sum of the kinetic energy and the potential energy: $E = KE + PE$.
- The work-energy theorem can be expressed in terms of the total mechanical energy as: $W_{nc} = E_f - E_0$, where W_{nc} is the net work done by the external nonconservative forces, and E_f and E_0 are the final and initial total mechanical energies, respectively.
- The **principle of conservation of mechanical**

energy states that the total mechanical energy E remains constant as the object moves, provided that the net work done by external nonconservative forces is zero. This follows from the work-energy theorem in the form $W_{nc} = E_f - E_0$. When the net work done by external nonconservative forces is zero, $W_{nc} = 0$, and

$$E_f = E_0 \qquad (6.9)$$

In other words, the mechanical energy remains constant at the value E_0 all along the path that connects the initial and final points.

- While the sum of the kinetic and potential energies at any point along the path is conserved, the two forms may be transformed into one another. For example, when a ball is thrown straight up in the absence of air resistance, the total mechanical energy is conserved. If we take $h = 0$ at the surface of the earth, the initial gravitational potential energy is zero, and the kinetic energy of the ball is equal to the total mechanical energy of the particle. As the ball rises, it slows down and its vertical height h from the surface of the earth increases. Therefore, the ball loses kinetic energy (since it is slowing down), and gains gravitational potential energy (since h is increasing). A simple calculation will show that the gain in gravitational potential energy between any two points on the path is equal to the loss in kinetic energy. At the ball's maximum height, $v = 0$, and its kinetic energy is zero. At that instant, all of the mechanical energy of the ball is in the form of gravitational potential energy; furthermore, the value of the ball's gravitational potential energy is equal to its original kinetic energy. Only the force of gravity (a conservative force) does work, so the total mechanical energy E remains constant at all points along the trajectory of the ball. As the ball falls from its maximum height, the ball loses gravitational potential energy and gains kinetic energy. Again, E remains constant at all points along the trajectory of the ball.

- *Reasoning Strategy*

Applying the Principle of Conservation of Mechanical Energy

> **Step 1:** Identify the external conservative and nonconservative forces that act on the object. In order for this principle to apply, any nonconservative forces must act perpendicular to the displacement of the object so they do no work: $W_{nc} = 0$.
>
> **Step 2:** Choose the zero level for the gravitational potential energy. While this location is arbitrary, it may not be

changed during the course of the problem.

Step 3: Set the final total mechanical energy of the object equal to the initial total mechanical energy, as in Equation 6.9.

6.7 POWER

- **Average power** $\overline{P}$ is the average rate at which work W is done. It is calculated as the work per unit time, or

$$\overline{P} = \frac{\text{Work}}{\text{Time}} = \frac{W}{t} \qquad (6.10)$$

- The SI unit of power is the watt (W).
  1 W = 1 J/s.
  The unit of power in the BE system is the foot·pound per second (ft·lb/s).
- The unit of horsepower (hp) is used to specify the power generated by electric motors and internal combustion engines:
  1 horsepower = 550 foot·pounds/second = 746 watts.

- As shown in the text, the average power can also be expressed as:

$$\overline{P} = F\overline{v} \qquad (6.11)$$

where $\overline{v}$ is the average speed of the object.

6.8 OTHER FORMS OF ENERGY AND THE CONSERVATION OF ENERGY

- There are many types of energy besides kinetic energy and gravitational potential energy. The text identifies electrical energy, heat energy, chemical energy, and nuclear energy. In general, all types of energy can be converted from one form to another.
- The **principle of conservation of energy** states that energy can neither be created nor destroyed, but can only be transformed from one form to another.

6.9 WORK DONE BY A VARIABLE FORCE

- In many situations, the force that moves an object between two points is not constant in magnitude. In these cases, the force is variable and one cannot calculate the work done by the force from the equation $W = (F \cos \theta) s$.
- The **work done by a variable force** in moving an object between two points is equal to the area under the graph of $(F \cos \theta)$ vs. s.
- Consider a sled that is pulled by a force **F** at an angle θ over a displacement **s**. The graph below shows how the magnitude of the quantity $(F \cos \theta)$ varies over the path. The area under the curve between any two values of s is equal to the work done between those two points. For example, the work done as the sled is pulled between the positions s_1 and s_2 is equal to the shaded area shown on the graph.

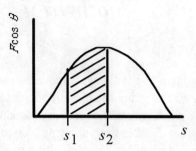

CHAPTER 7 | *IMPULSE AND MOMENTUM*

In principle, it is always possible to use Newton's laws to predict the motion of objects. In practice, however, unless the force is constant, or changes in a predictable way, applying Newton's second law to determine the acceleration of an object may be tedious at best, if not impossible without advanced computational methods.

In this chapter, we introduce the concepts of impulse and momentum. By employing Newton's third law to the collision between two objects, we obtain the principle of conservation of linear momentum. This conservation principle permits the analysis of the motion of interacting objects in situations in which it would be too difficult to apply Newton's second law directly.

Important Concepts

- impulse
- linear momentum
- impulse-momentum theorem
- internal forces
- external forces
- isolated system

- conservation of linear momentum
- elastic collision
- inelastic collision
- completely inelastic collision
- center of mass
- velocity of the center of mass

7.1 THE IMPULSE-MOMENTUM THEOREM

- The **impulse** of a force is the product of the average force $\overline{\mathbf{F}}$ and the time interval Δt during which the force acts:

$$\textbf{Impulse} = \overline{\mathbf{F}}\Delta t \qquad (7.1)$$

- Impulse is a vector that points in the same direction as the average force.
- The SI unit for impulse is the newton·second (N·s). Its dimension is $[M][L][T]^{-1}$.
- The **linear momentum p** of an object is the product of the object's mass m and velocity $\mathbf{v}$:

$$\mathbf{p} = m\mathbf{v} \qquad (7.2)$$

- Linear momentum is a vector that points in the same direction as the velocity.
- The SI unit for linear momentum is the kilogram·meter/second (kg·m/s). Its dimension is $[M][L][T]^{-1}$.
- The **impulse-momentum theorem** states that when a net force **F** acts on an object, the impulse of the net force is equal to the change in the momentum of the object:

$$\underbrace{\overline{\mathbf{F}}\Delta t}_{\text{Impulse}} = \underbrace{m\mathbf{v}_f}_{\substack{\text{Final} \\ \text{momentum}}} - \underbrace{m\mathbf{v}_0}_{\substack{\text{Initial} \\ \text{momentum}}} \qquad (7.4)$$

- The average force during an interaction is difficult to measure directly; thus, it is difficult to measure the impulse directly. The impulse-momentum theorem allows us to obtain information about the impulse indirectly by measuring the change in momentum that the impulse causes.
- Note that in order for Equation 7.4 to be a valid equation, the units and dimensions of impulse and linear momentum must be equal.

7.2 THE PRINCIPLE OF CONSERVATION OF LINEAR MOMENTUM

- Two types of forces arise during the collision of two or more objects:
  Internal forces — Forces that the objects within the system exert on each other.
  External forces — Forces exerted on the objects by agents that are external to the system.
- An **isolated system** is one for which the sum of the external forces acting on the system is zero.
- When the impulse momentum theorem and Newton's third law are applied to a collision of two or more objects, we find that

(Sum of average external forces) $\Delta t = \mathbf{P}_f - \mathbf{P}_0$
(7.6)

where $\mathbf{P}_0$ and $\mathbf{P}_f$ are the initial and final momenta of the system of objects. If this system of objects is isolated, the sum of the external forces is zero, and we obtain

$$\mathbf{P}_f = \mathbf{P}_o \qquad (7.7)$$

- Equation 7.7 is known as the **principle of conservation of linear momentum** and states that the total linear momentum of an isolated system remains constant (is conserved).
- *Reasoning Strategy*
  Applying the Principle of Conservation of Momentum
  Step 1: Select the object or objects (the "system") under consideration.
  Step 2: For the system chosen in Step 1, identify the internal forces and the external forces.
  Step 3: Verify that the system is isolated - that is, verify that the sum of the external forces acting on the system is zero. If the system is not isolated, choose a different system for analysis.
  Step 4: Keeping in mind that momentum is a vector, set the total final momentum of the system equal to the total initial momentum.

7.3 COLLISIONS IN ONE DIMENSION

- Collisions are classified according to whether the total kinetic energy is conserved during the collision:
  1. **Elastic collision** — One in which the total kinetic energy of the system after the collision is equal to the total kinetic energy before the collision.
  2. **Inelastic collision** — One in which the total kinetic energy of the system is *not* the same before and after the collision; if the objects stick together after colliding, the collision is said to be *completely inelastic*.
- When two objects with masses m_1 and m_2 respectively, initially traveling with speeds v_{01} and v_{02} collide *elastically* in the absence of a net external force, their respective velocities immediately after the collision are

$$v_{f1} = \left(\frac{m_1 - m_2}{m_1 + m_2}\right) v_{01} \quad (7.8a)$$

$$v_{f2} = \left(\frac{2m_1}{m_1 + m_2}\right) v_{01} \quad (7.8b)$$

7.4 COLLISIONS IN TWO DIMENSIONS

- Linear momentum is a vector, and in two dimensions the x and y components of the total linear momentum are conserved separately. Equation 7.7 is equivalent to the following two equations:

[**x component**] $\qquad P_{fx} = P_{0x} \qquad$ (7.9a)

[**y component**] $\qquad P_{fy} = P_{0y} \qquad$ (7.9b)

7.5 CENTER OF MASS

- The **center of mass** of a system of objects is a point that represents the average location for the total mass of the system.
- The location x_{cm} of the center of mass of two objects lying on the x axis is defined to be

$$x_{cm} = \frac{m_1 x_1 + m_2 x_2}{m_1 + m_2} \qquad (7.10)$$

where m_1 and m_2 are the masses of the objects, and x_1 and x_2 are their distances from the coordinate origin.

- Note that each term in the numerator of Equation 7.10 is the product of the particle's mass and position, while the denominator is the total mass of the system. Using this pattern, Equation 7.10 can be generalized for a system of any number of particles. It may also be generalized for particles that lie on the y or z axis.
- If the objects are moving with velocities v_1 and v_2, the velocity v_{cm} of the center of mass is given by

$$v_{cm} = \frac{m_1 v_1 + m_2 v_2}{m_1 + m_2} \quad (7.11)$$

- Equation 7.11 can also be generalized to accommodate a system of many objects.
- If the total linear momentum of a system of objects remains constant during an interaction such as a collision, the velocity of the center of mass also remains constant.

CHAPTER 8 | ROTATIONAL KINEMATICS

This chapter introduces the study of rotational motion. In particular, the discussion will focus on the rotation of a rigid body about a fixed axis. While it is possible to describe such motion using the kinematic variables introduced in Chapter 2, these variables do not provide the most efficient description of rotational motion.

The approach followed by the text is to introduce and define new kinematic variables that efficiently describe rotational motion, and then show how these new kinematic variables are related to the variables introduced for linear motion in chapter 2.

Important Concepts

- angular displacement
- degree
- revolution
- radian
- average angular velocity
- instantaneous angular velocity
- instantaneous angular speed
- average angular acceleration

- instantaneous angular acceleration
- equations of rotational kinematics
- tangential speed
- tangential acceleration
- centripetal acceleration
- rolling motion
- linear speed
- linear acceleration
- right-hand rule

8.1 ROTATIONAL MOTION AND ANGULAR DISPLACEMENT

- When a rigid body rotates about a fixed axis, the **angular displacement** is the angle $\Delta\theta$ swept out by a line passing through any point on the body and intersecting the axis of rotation perpendicularly.
- By convention, the angular displacement is positive if it is counterclockwise and negative if it is clockwise.
- The SI unit of angular displacement is the radian (rad). The radian is neither a base nor a derived SI unit. It is considered to be a supplementary SI unit.

- The angle θ in radians is defined from the following figure:

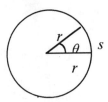

$$\theta \text{ (in radians)} = \frac{\text{Arc length}}{\text{Radius}} = \frac{s}{r} \qquad (8.1)$$

Note that Equation 8.1 is only valid when the angle θ is expressed in radians. Since θ is the ratio of two lengths, it is a *dimensionless* or pure number.
- Angular displacement can also be expressed in two other units, the *degree* and the *revolution* (*rev*). There are 360° in 1 revolution.
- From Equation 8.1, it follows that 2π rad = 360° = 1 rev.

8.2 ANGULAR VELOCITY AND ANGULAR ACCELERATION

- For rotational motion about a fixed axis, the **average angular velocity** is defined by

$$\overline{\omega} = \frac{\theta - \theta_o}{t - t_o} = \frac{\Delta\theta}{\Delta t} \quad (8.2)$$

- The SI unit and dimension of angular velocity are, respectively, the radian per second (rad/s) and [1/T].
- Average angular velocity is a vector quantity. The direction of the average angular velocity is the same as that of the angular displacement.
- The **instantaneous angular velocity,** or the angular velocity at any given instant, is defined by

$$\omega = \lim_{\Delta t \to 0} \overline{\omega} = \lim_{\Delta t \to 0} \frac{\Delta\theta}{\Delta t} \quad (8.3)$$

- The magnitude of the instantaneous angular velocity is called the **instantaneous angular speed**.

- When the angular velocity changes from an initial value of ω_o at time t_o to a later value of ω at time t, the **average angular acceleration** is defined by

$$\bar{\alpha} = \frac{\Delta \omega}{\Delta t} \qquad (8.4)$$

- The SI unit and dimension for average angular acceleration are, respectively, radians per second per second (rad/s^2) and [1/T^2].
- Average angular velocity is a vector quantity. Its direction is the same as the direction of the *change* in the angular velocity.
- When Δt approaches zero, the average angular acceleration becomes equal to the angular acceleration at a particular instant or the **instantaneous angular acceleration** α.

8.3 THE EQUATIONS OF ROTATIONAL KINEMATICS

- Suppose that a rigid body rotates about a fixed axis with constant angular acceleration α, and at time $t = 0$, $\theta_0 = 0$, so that the angular displacement is $\Delta\theta = \theta - \theta_o = \theta$. Then, the angular displacement θ, the final angular velocity ω, the initial angular velocity ω_0, and the elapsed time t are related by **the equations of rotational kinematics** given in the left column of the table below. The table also shows the analogous relation for linear motion with constant acceleration **a**.

Rotational Motion (α = constant)		Linear Motion (a = constant)	
$\omega = \omega_0 + \alpha t$	(8.4)	$v = v_0 + at$	(2.4)
$\theta = \frac{1}{2}(\omega_0 + \omega)t$	(8.6)	$x = \frac{1}{2}(v_0 + v)t$	(2.7)
$\theta = \omega_0 t + \frac{1}{2}\alpha t^2$	(8.7)	$x = v_0 t + \frac{1}{2}at^2$	(2.8)
$\omega^2 = \omega_0^2 + 2\alpha\theta$	(8.8)	$v^2 = v_0^2 + 2ax$	(2.9)

- The equations for rotational kinematics may be used with any self-consistent set of units and are not restricted to radian measure.

8.4 ANGULAR VARIABLES AND TANGENTIAL VARIABLES

- When a rigid body rotates through an angle θ about a fixed axis, any point on the body moves on a circular arc of length s and radius r. Such a point has a tangential velocity $\mathbf{v}_T$, and if the magnitude of $\mathbf{v}_T$ is changing, it has a tangential acceleration $\mathbf{a}_T$.

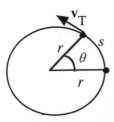

- The angular and tangential variables are related according to the following equations:

$$s = r\theta \quad (\theta \text{ in radians}) \qquad (8.1)$$

$$v_T = r\omega \quad (\omega \text{ in rad/s}) \qquad (8.9)$$

$$a_T = r\alpha \quad (\alpha \text{ in rad/s}^2) \qquad (8.10)$$

- These equations make reference to the magnitudes of the variables involved. They make no reference to direction and should *not* include positive or negative signs.
- These equations are only valid when the angular quantities are expressed in radians.

8.5 CENTRIPETAL ACCELERATION AND TANGENTIAL ACCELERATION

- When an object speeds up as it moves in a circle about a fixed axis, it has a tangential acceleration, as discussed in the last section. Furthermore, the object also has a centripetal acceleration as described in Chapter 5.
- The magnitude of the centripetal acceleration can be expressed in terms of the angular speed ω and is given by

$$a_c = r\omega^2 \qquad (8.11)$$

- A point on an object rotating with **nonuniform circular motion** experiences a total acceleration that is the vector sum of two perpendicular acceleration components, the centripetal acceleration $\mathbf{a}_c$ and the tangential acceleration $\mathbf{a}_T$:

$\mathbf{a} = \mathbf{a}_c + \mathbf{a}_T$.

Since $\mathbf{a}_c$ and $\mathbf{a}_T$ are perpendicular, the magnitude of **a** can be found by using Pythagorean's theorem:

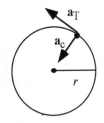

$$a = \sqrt{a_c^2 + a_T^2}.$$

8.6 ROLLING MOTION

- In **rolling motion**, there is no slipping at the point where the object touches the surface upon which it is rolling.
- The tangential speed v_T of a point on the outer edge of the rolling object, measured relative to the axis through the center of the object, is equal to the linear speed v with which the object moves parallel to the surface.
- The linear speed with which the object moves parallel to the surface is related to the angular speed ω about the rotation axis according to

$$\underbrace{v}_{\text{Linear speed}} = \underbrace{r\omega}_{\text{Tangential speed, } v_T}. \quad (\omega \text{ in rad/s}) \quad (8.12)$$

- Similarly, the magnitude of the linear acceleration a of a rolling object is related to α, the magnitude of the angular acceleration according to

$$\underbrace{a}_{\text{Linear acceleration}} = \underbrace{r\alpha}_{\text{Tangential acceleration, } a_T}. \quad (\alpha \text{ in rad/s}^2)$$

(8.13)

*8.7 THE VECTOR NATURE OF ANGULAR VARIABLES

- The angular displacement $\Delta\theta$, the angular velocity ω, and the angular acceleration α, are vector quantities.
- The direction of the angular velocity can be found by employing the **right-hand rule**:
  Right-Hand Rule: Imagine grasping the axis of rotation with your right hand, so that your fingers circle the axis in the same sense as the rotation. Your extended thumb points along the

axis in the direction of the angular velocity vector.

- Since the angular velocity has the same direction as the angular displacement, the right-hand rule can also be used to determine the direction of the angular displacement.
- The angular acceleration vector has the same direction as the *change* in the angular velocity. Thus, when the angular velocity is constant, the angular acceleration is zero. When the angular velocity is increasing, the angular acceleration points in the same direction as the angular velocity; when the angular velocity is decreasing, the angular acceleration points in the direction opposite to the angular velocity.

CHAPTER 9 | *ROTATIONAL DYNAMICS*

Chapter 8 introduced the descriptive nature of angular motion, namely *rotational kinematics*. This chapter introduces *rotational dynamics* — the study of the causes of rotational motion. Rotational kinematics examines the relationships between the rotational kinematic variables (θ, ω, α) and the properties of the body or system undergoing rotational motion.

In this chapter, you will be introduced to the concept of torque, which is the rotational analogue of force, and to angular momentum, which is the rotational analogue of linear momentum. You will see that for systems in which the net external torque is zero, the angular momentum of the system is conserved. Like the principle of conservation of linear momentum, the principle of conservation of angular momentum is a powerful calculational tool.

Important Concepts

- rotation axis
- line of action
- lever arm
- torque
- equilibrium of a rigid object
- center of gravity

- moment of inertia
- Newton's second law for rotational motion
- rotational work
- rotational kinetic energy
- angular momentum
- principle of conservation of angular momentum

9.1 THE EFFECTS OF FORCES AND TORQUES ON THE MOTION OF RIGID OBJECTS

- The **line of action** of a force **F** is an extended line drawn colinear with the force.
- The **lever arm** ℓ of the force **F** is the distance between the line of action and the axis of rotation, measured on a line that is perpendicular to both.
- The **torque** τ of the force **F** has a magnitude that is given by

$$\tau = F\ell. \qquad (9.1)$$

- Torque is a vector quantity. While the direction of the torque can be specified by the *right-hand-rule* introduced in Chapter 8, its direction is more commonly given with plus and minus signs. It is positive if the force tends to produce a counterclockwise rotation about the axis, and negative if the force tends to produce a clockwise rotation.

9.2 RIGID OBJECTS IN EQUILIBRIUM

- When a rigid body is in **equilibrium**, its linear and rotational motion do not change; such an object has no acceleration of any kind — the object has zero translational acceleration and zero angular acceleration.
- For a rigid body in equilibrium, the sum of the externally applied forces is zero and the sum of the externally applied torques is zero:

$$\sum F_x = 0 \quad \text{and} \quad \sum F_y = 0 \quad (4.9\text{a and } 4.9\text{b})$$
$$\sum \tau = 0 \quad (9.2)$$

- For a rigid body in equilibrium, Equation 9.2 must hold for *any arbitrarily chosen axis of rotation. If an object is in equilibrium, it is in equilibrium with respect to any axis whatsoever.*

9.3 CENTER OF GRAVITY

- The **center of gravity** of a rigid body is the point at which its weight *can be considered to act* when calculating the torque due to the weight of the object.
- For a symmetrical body with uniformly distributed weight, the center of gravity is located at the geometrical center of the body.
- When N objects, whose weights are $W_1, W_2, \ldots W_N$, are distributed along the x axis at locations $x_1, x_2, \ldots x_N$, the center of gravity is located at a point whose x coordinate is given by

$$x_{cg} = \frac{W_1 x_1 + W_2 x_2 + \ldots + W_N x_N}{W_1 + W_2 + \ldots + W_N}. \qquad (9.3)$$

- The location of the *center of mass* is identical to the location of the center of gravity, provided that the acceleration due to gravity does not vary over the physical extent of the objects in the system.

9.4 NEWTON'S SECOND LAW FOR ROTATIONAL MOTION ABOUT A FIXED AXIS

- For a single particle of mass m that rotates about a fixed axis in a circle of radius r, Newton's second law takes the form

$$\tau = \underbrace{(mr^2)}_{\text{Moment of Inertia, } I} \alpha \qquad (9.4)$$

where the proportionality constant $I = mr^2$ is the **moment of inertia** of the particle relative to the rotation axis.

- The **moment of inertia** I of an object composed of N particles is $I = (m_1 r_1^2 + m_2 r_2^2 + \ldots + m_N r_N^2)$ where $m_1, m_2, \ldots m_N$ are the masses of the particles, and $r_1, r_2, \ldots r_N$ are the perpendicular distances of the particles from the axis of rotation. This equation can be written in the shorthand notation:

$$I = \sum mr^2 \qquad (9.6)$$

- The moment of inertia of a rigid object depends on the location and orientation of the axis relative to the particles that make up the object. *The moment of inertia is, for the same object, different for different rotation axes.*
- Table 9.1 in the text gives the moments of inertia for some common objects.
- For rigid bodies rotating about fixed axes, **Newton's second law for rotational motion** is

$$\sum \tau = I\alpha \qquad (9.7)$$

where I is the moment of inertia of the body and α is the angular acceleration expressed in rad/s^2.
- The direction of the net torque on a rigid body is the same as the direction of the angular acceleration.

9.5 ROTATIONAL WORK AND ENERGY

- The **rotational work** W_R done by a constant torque τ in turning a rigid body through an angular displacement θ (in radians) is given by

$$W_R = \tau\theta \qquad (9.8)$$

- Equation 9.8 is only valid when the angle θ is expressed in radians. Equation 9.8 then gives the usual SI unit for work, the joule (J).
- The **rotational kinetic energy** KE_R of a rigid object rotating with an angular speed ω about a fixed axis and having a moment of inertia I is

$$KE_R = \frac{1}{2} I \omega^2 \qquad (9.9)$$

- Often the motion of an object is a combination of translational and rotational motion. For example, when a car or a bicycle is in motion, its wheels are both translating and rotating.
- The total kinetic energy KE of an object that is simultaneously translating and rotating is the sum of its translational and rotational kinetic energies.

9.6 ANGULAR MOMENTUM

- The **angular momentum** L of a body rotating with an angular velocity ω (in rad/s) about a fixed axis and having a moment of inertia I about that axis is

$$L = I\omega \qquad (9.10)$$

- The SI unit and dimension of angular momentum are, respectively, $kg \cdot m^2 / s^2$ and $[M][L]^2[T]^{-2}$.
- Angular momentum is a vector. It points in the same direction as the angular velocity.
- The **principle of conservation of angular momentum** states that the total angular momentum of a system remains constant (is conserved) if the net external torque acting on the system is zero.
- The conservation of angular momentum can be used to show that a satellite does not have a constant speed when it travels in an elliptical orbit.

CHAPTER 10 | *ELASTICITY AND SIMPLE HARMONIC MOTION*

This chapter discusses the elastic properties of materials and the kinds of deformations a material can undergo when subjected to external forces. As a special case, the deformation of a spring and the resulting motion will be discussed in detail. This important kind of oscillatory or vibrational motion is called *simple harmonic motion*. You will see that an object attached to a spring, retraces its motion after a fixed period of time and continues to do so at regular intervals. Furthermore, you will see that simple harmonic motion has many applications in the real world.

Important Concepts

- elastic deformation
- stretch
- compression
- Young's modulus
- shear deformation
- shear modulus
- volume deformation

- bulk modulus
- stress
- strain
- Hooke's law
- ideal spring
- restoring force
- simple harmonic motion
- reference circle
- angular frequency
- elastic potential energy
- simple pendulum
- damped harmonic motion
- driven harmonic motion
- resonance

10.1 ELASTIC DEFORMATION

- All materials become distorted when they are stretched or compressed. Within a certain limit (called the *elastic limit* of the material), many materials exhibit *elastic* behavior and return to their original shape when the stretching or compression is removed.

- The possible kinds of **elastic deformation** include *stretch and compression, shear deformation*, and v*olume deformation.*
- The magnitude F of the force required to stretch or compress an object of original length L_o and cross-sectional area A by an amount ΔL is given by

$$F = Y \frac{\Delta L}{L_o} A \qquad (10.1)$$

The term Y is a proportionality constant called **Young's modulus** and has units of force per unit area (N/m^2). Table 10.1 in the text gives some representative values for the Young's modulus of various solids.

- Many solids have relatively large Young's moduli, reflecting the fact that a large force is required to change the length of a solid object by even a small amount.
- Note that the magnitude of the force in Equation 10.1 is *proportional to the fractional change in length* ($\Delta L / L_o$), rather than the absolute change in length ΔL.
- The magnitude F of the force required to produce an amount of shear ΔX for an object with thickness L_o and cross-sectional area A is given by

$$F = S\left(\frac{\Delta X}{L_o}\right) A \qquad (10.2)$$

where the constant of proportionality S is called the **shear modulus** and, like Young's modulus, has units of force per unit area (N/m^2). The value of the shear modulus depends on the material. Table 10.2 in the text gives some representative values for the shear moduli of various materials.

- The **pressure** P exerted on a surface by a force **F** is the magnitude F of the force acting perpendicular to a surface divided by the area A over which the force acts:

$$P = \frac{F}{A} \qquad (10.3)$$

- Pressure is a scalar quantity and has no associated direction.
- The SI unit of pressure is the pascal (Pa) where 1 Pa = 1 N/m^2. The dimension of pressure is $[M][L]^{-1}[T]^{-2}$.
- The result of increasing the pressure on an object by an amount ΔP is that the volume of the object decreases by an amount ΔV. The amount of pressure increase ΔP needed to decrease the volume of an object of original volume V_o by an amount ΔV is given by

$$\Delta P = -B \left(\frac{\Delta V}{V_o} \right) \qquad (10.4)$$

where the proportionality constant B is called the **bulk modulus**. Like Young's modulus and the shear modulus, the bulk modulus has units of force per unit area (N/m^2). Table 10.3 in the text gives values for the bulk moduli of some representative liquids and solids.

10.2 STRESS, STRAIN, AND HOOKE'S LAW

- **Stress** is the force per unit area applied to an object and causes **strain**.
- For *stretch and compression* and *volume deformation*, strain is the resulting fractional change in length or volume. For *shear deformation*, strain reflects the change in the shape of the object.
- **Hooke's law** states that stress is directly proportional to strain, up to a limit called the **proportionality limit**.

10.3 THE IDEAL SPRING AND SIMPLE HARMONIC MOTION

- For relatively small deformations, the force **F** that must be applied to stretch or compress an **ideal spring** by a displacement x from its unstrained length is

$$F = kx \qquad (10.5)$$

where the proportionality constant k is called the **spring constant**.

- A spring exerts a **restoring force** on an object attached to the spring. The restoring force produced by an ideal spring is given by

$$F = -kx \qquad (10.6)$$

where the minus sign indicates that the restoring force points opposite to the displacement of the spring.

- The **equilibrium position** of an object attached to a horizontal spring corresponds to the location of the object when the spring exhibits its unstrained length (that is, the equilibrium position is the position where the spring is neither stretched nor compressed). At the equilibrium position, neither the object nor the spring exert forces on each other.

- **Simple harmonic motion** is the oscillatory motion that occurs when a restoring force of the form of Equation (10.6) acts on an object. The object continually retraces a linear path that is symmetrically oriented about its equilibrium position. From Equation 10.6 and Newton's second law, we see that the object moves over this path with an acceleration that is directly proportional to the displacement from equilibrium and oriented so that its direction is always opposite to the displacement.

- When an object attached to a horizontal spring is moved from its equilibrium position and released, the restoring force $F = -kx$ leads to simple harmonic motion. When an object attached to a vertical spring is moved from its equilibrium position and released, the resulting motion is also simple harmonic motion. However, when the spring is vertical, the weight of the object causes the spring to stretch, and the equilibrium position is different from the case in which the spring is horizontal.

10.4 SIMPLE HARMONIC MOTION AND THE REFERENCE CIRCLE

- The projection of uniform circular motion along any diameter of a circular path (known as a **reference circle**) is a good model for simple harmonic motion. The text uses this relationship between uniform circular motion and simple harmonic motion to develop a mathematical description for simple harmonic motion.
- The **amplitude** A of the motion is the maximum distance that the object moves away from its equilibrium position.
- The **period** T of the motion is the time required for the object to complete one cycle of the motion.
- The displacement x of the object from its equilibrium position, as a function of time t, is given by

$$x = A\cos\theta = A\cos\omega t \qquad (10.7)$$

where ω, the **angular frequency** of the motion (in rad/s), is given by

$$\omega = \frac{2\pi}{T}. \qquad (10.8)$$

- The **frequency** f of the motion is the number of cycles per second that occurs. Frequency and period are related according to

$$f = \frac{1}{T} \qquad (10.9)$$

and, consequently, the frequency f and the angular frequency ω are related according to

$$\omega = \frac{2\pi}{T} = 2\pi f. \qquad (10.10)$$

- The velocity in simple harmonic motion, as a function of the time t, is given by

$$v = -A\omega \sin \theta = -A\omega \sin \omega t. \qquad (10.11)$$

- When the object passes through its equilibrium position, its velocity has its maximum magnitude given by

$$v_{max} = A\omega \qquad (\omega \text{ in rad}/\text{s}). \qquad (10.12)$$

Conversely, when the object is at its maximum displacement from the equilibrium position (that is, when $x = \pm A$), its velocity is zero.
- The acceleration in simple harmonic motion, as a function of the time t, is given by

$$a = -A\omega^2 \cos\theta = -A\omega^2 \cos \omega t \qquad (10.13)$$

- When the object passes through its equilibrium position, its acceleration is zero. When the object has reached its maximum displacement from the equilibrium position (that is, when $x = \pm A$), the acceleration has its maximum magnitude given by

$$a_{max} = A\omega^2 \qquad (\omega \text{ in rad / s}) \qquad (10.14)$$

- Note that, in simple harmonic motion, both the velocity and acceleration are continually changing with time. Furthermore, when the magnitude of the velocity is a maximum, the acceleration is zero, and vice versa.
- For an object of mass m on a spring with spring constant k, the angular frequency ω and the frequency f are determined by

$$\omega = 2\pi f = \sqrt{k/m} \quad (\omega \text{ in rad/s}) \qquad (10.15)$$

10.5 ENERGY AND SIMPLE HARMONIC MOTION

- When a spring is stretched or compressed, it has potential energy called **elastic potential energy**. Consequently, a stretched or compressed spring can do work on an object that is attached to the spring.
- When an object that is attached to a spring is moved from an initial position x_o to a final position x_f, the work done by the spring force is given by

$$W_{\text{elastic}} = \underbrace{\frac{1}{2}kx_0^2}_{\text{Initial elastic potential energy}} - \underbrace{\frac{1}{2}kx_f^2}_{\text{Final elastic potential energy}}. \qquad (10.16)$$

- The **elastic potential energy** of an object attached to an ideal spring that has a spring constant k and is stretched or compressed by an amount x relative to its unstrained length is

$$\text{PE}_{\text{elastic}} = \frac{1}{2}kx^2. \qquad (10.17)$$

- The total mechanical energy E of a system composed of an object attached to an ideal spring is the sum of its translational and rotational kinetic energies, its gravitational potential energy, and its elastic potential energy:

$$E = \frac{1}{2}mv^2 + \frac{1}{2}I\omega^2 + mgh + \frac{1}{2}kx^2 \quad (10.18)$$

- If there is no net work done by external nonconservative forces (like friction), then the total mechanical energy of the system is conserved, $E_f = E_o$.

10.6 THE PENDULUM

- A **simple pendulum** is a particle of mass m attached to a frictionless pivot by a cable whose length is L and whose mass is negligible.
- The small-angle ($\leq 10°$) back-and-forth swinging of a simple pendulum is simple harmonic motion, while large-angle motion is not.
- The angular frequency ω and the frequency f of small-angle motion are given by

$$\omega = 2\pi f = \sqrt{\frac{g}{L}} \quad \text{(small angles only).} \quad (10.20)$$

- A **physical pendulum** consists of a rigid object, with moment of inertia I and mass m, suspended from a frictionless pivot.
- For small oscillations of a physical pendulum, the motion is approximately simple harmonic motion with frequency

$$\omega = 2\pi f = \sqrt{\frac{mgL}{I}}$$

where L is the distance between the axis of rotation and the center of gravity of the rigid object.

10.7 DAMPED HARMONIC MOTION

- An object executing simple harmonic motion oscillates with a constant amplitude because there is no mechanism for energy dissipation.
- When friction or some other energy-dissipating mechanism is present, the amplitude decreases with time and the motion is no longer simple harmonic motion.

- **Damped harmonic motion** is motion in which the amplitude of oscillation decreases as time passes.
- A shock absorber in the suspension system of an automobile is designed to introduce damping forces which reduce vibrations.
- **Critical damping** is the minimum degree of damping that eliminates any oscillations in the motion as the object returns to its equilibrium position. When the damping exceeds its critical value, the motion is said to be *overdamped*. Conversely, when the damping is less than the critical value, the motion is said to be *underdamped*.

10.8 DRIVEN HARMONIC MOTION AND RESONANCE

- When sufficient energy is continually added to an oscillating system, the amplitude of the motion increases with time.
- **Driven harmonic motion** occurs when an additional driving force is applied to an object along with the restoring force. The additional driving force drives or controls the behavior of the system to a large extent.

- The *natural frequency* of driven harmonic motion is given by Equation (10.15):
  $f = (1/2\pi)\sqrt{k/m}$.
- **Resonance** is the condition under which a driving force can transmit large amounts of energy to an oscillating object, leading to large-amplitude motion. In the absence of damping, resonance occurs when the frequency of the driving force is equal to the natural frequency at which the object oscillates.

CHAPTER 11 | *FLUIDS*

This chapter introduces the study of fluids, including both static fluids and fluids in motion. It is possible to describe properties of static fluids and basic concepts of fluid flow in terms of concepts presented in earlier chapters, such as mass, velocity, force, and energy.

Material is presented on such properties of fluids as mass density, pressure, and Archimedes' principle. Basic concepts of fluid flow, such as Bernoulli's equation and viscosity, are also introduced.

Understanding of concepts from previous chapters such as force, work, and energy are essential to this chapter.

Important Concepts

- mass density
- specific gravity
- pressure
- pascal
- bar
- *psi*
- atmospheric pressure at sea level
- atmosphere
- pressure increment

- incompressible fluid
- mercury barometer
- open-tube manometer
- gauge pressure
- absolute pressure
- sphygmomanometer
- systolic pressure
- diastolic pressure
- Pascal's principle
- buoyant force
- Archimedes' principle
- fluid flow
- steady fluid flow
- unsteady fluid flow
- turbulent flow
- compressible fluid flow
- incompressible fluid flow
- viscous fluid flow
- nonviscous fluid flow
- ideal fluid
- rotational fluid flow
- irrotational fluid flow
- streamlines
- equation of continuity
- mass flow rate
- volume flow rate
- Bernoulli's equation
- lift
- Torricelli's equation

- laminar flow
- coefficient of viscosity
- poise
- Poiseuille's Law

11.1 MASS DENSITY

- The **mass density** is the mass m of a substance divided by its volume V:

$$\rho = \frac{m}{V} \qquad (11.1)$$

- The SI unit for mass density is kilogram/meter^3. Its dimension is $[M]/[L]^3$.
- Gases have the smallest densities, because gas molecules are relatively far apart and a gas contains a large fraction of empty space.
- The density of a substance depends on temperature and pressure.
- The **specific gravity** of a substance is its density divided by the density of a reference material, usually chosen to be water at 4°C (1.000×10^3 kg/m^3).

$$\text{Specific gravity} = \frac{\text{Density of substance}}{\text{Density of water at 4°C}} = \frac{\text{Density of substance}}{(1.000 \times 10^3 \text{ kg/m}^3)} \quad (11.2)$$

- Specific gravity has no units.

11.2 Pressure

- The **pressure** P exerted by a fluid is defined as the magnitude F of the force acting perpendicular to a surface divided by the area over which the force acts:

$$P = \frac{F}{A} \quad (11.3)$$

- The SI unit for pressure is a newton/meter^2 (N/m^2) which is referred to as a **pascal** (Pa).
- A **bar** is a pressure of 10^5 Pa.
- A unit of pressure is pounds per square inch (lb/in^2), abbreviated as "***psi***."
- A static fluid cannot produce a force parallel to a surface.

- Pressure is not a vector quantity. In the definition of pressure, $P = F/A$, the symbol F refers only to the magnitude of the force.
- Pressure has no directional characteristics.
  Atmospheric pressure at sea level is defined as 1.013×10^5 Pa = 1 atmosphere (atm).
  One atmosphere corresponds to 14.70 lb/in^2.

11.3 PRESSURE AND DEPTH IN A STATIC FLUID

- There are two types of external forces that act on a sample of fluid.
  Gravitational force is the weight of the object.
  Collisional force is due to molecular collisions and is responsible for pressure.
- For a column of fluid that is in equilibrium the sum of the vertical forces equals zero:

$$\sum F_1 = P_2 A - P_1 A - mg \text{ or } P_2 A = P_1 A + mg$$

where P_1 generates a downward force at the top of the column, P_2 generates the upward force at the bottom of the column, A is the area at the top (and bottom) of the column, and mg is the weight of the fluid in the column. This can be expressed as

$$P_2 = P_1 + \rho g h \qquad (11.4)$$

after division by the volume of the fluid in the column, V.

- $\rho g h$ is called the **pressure increment**.
- An **incompressible fluid** is a fluid that has the same density, ρ, at all depths. This is a reasonable assumption to make for liquids.
- The pressure increment $\rho g h$ is affected by the vertical distance h, but not the horizontal distance within the fluid.

11.4 PRESSURE GAUGES

- A **mercury barometer** is a tube sealed at one end, filled completely with mercury, and then inverted, so that the open end is under the

surface of a pool of mercury. This is a device used as a pressure gauge.
- The **open-tube manometer** is an U-tube containing mercury open to the atmosphere at one end with the other end connected to the container whose pressure is to be measured.
- The **gauge pressure** is the amount by which a container pressure differs from atmospheric pressure.
- **Absolute pressure** is actual pressure.
- The **sphygmomanometer** is a device used for measuring blood pressure, consisting of a cuff and a bulb.
- The **systolic pressure** is the pressure created by the heart at the peak of its beating cycle.
- The **diastolic pressure** is the pressure created by the heart at the low point of its beating cycle.

11.5 PASCAL'S PRINCIPLE

- **Pascal's principle**:
 Any change in the pressure applied to a completely enclosed fluid is transmitted undiminished to all parts of the fluid and the enclosing walls.
 For an enclosed fluid, force F_1 applied to one cap of area A_1, due to pressure P_1, must be

related to the force F_2 applied to the other cap of area A_2, due to pressure P_2, since $P_1 = P_2$ when the heights of the caps are equal.

$$F_2 = F_1\left(\frac{A_2}{A_1}\right) \qquad (11.5)$$

In a device such as a hydraulic car lift, the same amount of work is done by both the input and output forces in the absence of friction. The cap with the larger area moves the smaller distance.

11.6 ARCHIMEDES' PRINCIPLE

- The upward force applied by a fluid is called the **buoyant force**.
- A liquid applies a net upward force, or buoyant force, on an object with the same horizontal area at the top (point 1) and the bottom (point 2):

$$F_B = P_2 A - P_1 A = (P_2 - P_1)A$$

- **Archimedes' principle**
  Any fluid applies a buoyant force to an object that is partially or completely immersed in it; the magnitude of the buoyant force equals the weight of the fluid that the object displaces:

$$F_B = W_{fluid} \qquad (11.6)$$

  F_B is the magnitude of the buoyant force and W_{fluid} is the weight of the displaced fluid. If the buoyant force is strong enough to balance the force of gravity, objects will float in the fluid. If the buoyant force is not sufficiently large to balance weight, even with the object completely submerged, the object will sink.
- Any object that is *solid throughout* will float in a liquid if the density of the object is less than or equal to the density of the liquid.

11.7 FLUIDS IN MOTION

- **Fluid flow** can be **steady** or **unsteady**.
- In **steady flow** the velocity of the fluid particles at any point is constant as time passes.

- In **unsteady flow** the velocity at a point in the fluid changes as time passes.
- **Turbulent flow** is an extreme kind of unsteady flow.
- Fluid flow can be **compressible** or **incompressible.**
- A fluid is **incompressible** if its density remains constant as pressure changes.
- Fluid flow can be **viscous** or **nonviscous.**
- A **viscous** fluid does not flow readily and is said to have a large viscosity.
- The flow of a viscous fluid is an energy-dissipating process.
- An incompressible, nonviscous fluid is called an **ideal fluid.**
- Fluid flow can be **rotational** or **irrotational**.
- The flow is **rotational** when a part of the fluid has rotational as well as translational motion.
- **Streamlines** are used to represent trajectories of the fluid particles.
- Steady flow is often called **streamline flow.**

11.8 THE EQUATION OF CONTINUITY

- The **equation of continuity** expresses the simple idea that a same quantity of a fluid passes every point in a closed pipe over the same time interval.
- The mass of fluid per second that flows through a tube is called the **mass flow rate.**
- Equation of Continuity
 The mass flow rate ($\rho A v$) has the same value at every position along a tube that has a single entry and a single exit point for fluid flow. For two positions along such a tube

$$\rho_1 A_1 v_1 = \rho_2 A_2 v_2 \qquad (11.8)$$

- The SI unit of Mass Flow Rate is kilograms/second.
- Its dimension is [M]/[T].
- For an incompressible fluid the equation of continuity reduces to

$$A_1 v_1 = A_2 v_2 \qquad (11.9)$$

- The volume of fluid per second that passes through a tube is the **Volume flow rate** Q:

$$Q = \text{volume flow rate} = Av \qquad (11.10)$$

11.9 BERNOULLI'S EQUATION

- **Bernoulli's equation**:
  In a steady, irrotational flow of a nonviscous, incompressible fluid of density ρ, the pressure P, the fluid speed v, and the elevation y at any two points (1 and 2) are related by

$$P_1 + \frac{1}{2}\rho v_1^2 + \rho g y_1 = P_2 + \frac{1}{2}\rho v_2^2 + \rho g y_2 \quad (11.11)$$

- Bernoulli's equation is derived from work-energy considerations.

11.10 APPLICATIONS OF BERNOULLI'S EQUATION

- When a moving fluid is constrained to a horizontal pipe, all parts of it have the same elevation ($y_1 = y_2$), and Bernoulli's equation simplifies to

$$P_1 + \frac{1}{2}\rho v_1^2 = P_2 + \frac{1}{2}\rho v_2^2 \quad (11.12)$$

- Bernoulli's equation can be used to determine **lift**. This occurs when the pressure above an object is less than the pressure below an object.
- **Torricelli's theorem**:
  The speed at which an ideal liquid leaves a pipe is the same as if the liquid had freely fallen a distance h equal to the distance from the pipe's exit to the top of the liquid, or

$$v = \sqrt{2gh}$$

11.11 VISCOUS FLOW

- The flow of a liquid considered to be broken into thin, parallel plates is called **laminar flow**.
- Force is needed to move a layer of viscous fluid with a constant velocity.
- The tangential force **F** required to move a fluid layer at a constant speed v, when the layer has an area A and is located a perpendicular distance y from an immobile surface, is given by

$$F = \frac{\eta A v}{y} \qquad (11.13)$$

where η is the **coefficient of viscosity**.
- The SI unit of viscosity is (pascal)(second).
- The common unit of viscosity is the poise (P).
  1 poise (P) = 0.1 Pa · s

- Under ordinary conditions the viscosities of gases are significantly smaller than those of liquids.
- The viscosities of either liquids or gases depend markedly on temperature.
  (Usually, the viscosities of liquids decrease as the temperature is increased).
- **Poiseuille's Law**:
  A fluid whose viscosity is η, flowing through a pipe of radius R and length L, has a volume flow rate Q given by

  $$Q = \frac{\pi R^4 (P_2 - P_1)}{8\eta L} \qquad (11.14)$$

  where P_1 and P_2 are the pressures at the ends of the pipe.

CHAPTER 12

TEMPERATURE AND HEAT

This chapter introduces heat as a form of energy and the concept of temperature.

The material contained in this chapter includes the states of matter and the phase changes that occur. Thermal effects on matter, such as thermal expansion are also presented. The concept of specific heat is also discussed.

Important Concepts

- Celsius scale
- Fahrenheit scale
- ice point
- steam point
- Kelvin temperature scale
- absolute zero
- constant-volume gas thermometer
- absolute zero point
- thermometric property
- thermocouple
- electrical resistance thermometer
- thermograph or thermogram
- linear expansion
- coefficient of linear expansion
- bimetallic strip
- coefficient of volume expansion

- heat
- specific heat capacity
- kilocalorie
- calorie
- British thermal unit
- mechanical equivalent of heat
- calorimeter
- calorimetry
- phase
- melt
- fuse
- freeze
- evaporate
- condense
- boiling
- sublime
- latent heat of fusion
- latent heat of vaporization
- latent heat of sublimation
- equilibrium vapor pressure
- vapor pressure curve
- vaporization curve
- partial pressure
- relative humidity
- saturated
- unsaturated
- dew point

12.1 COMMON TEMPERATURE SCALES

- **Thermometers** are used to measure **temperature**. Many thermometers make use of the fact that materials usually expand when their temperatures increase.
- The temperature at which ice melts under one atmosphere of pressure is the **ice point**.
- The temperature at which water boils under one atmosphere of pressure is the **steam point**.
- Temperature scales are defined by assigning two temperature points on the scale and dividing the distance between them into a number of equally spaced intervals.
- The **Celsius** scale uses an ice point of 0°C and a steam point of 100°C.
- The **Fahrenheit** scale uses an ice point of 32°F and a steam point of 212°F.
- There is a subtle difference in the way the temperature of an object and the change of its temperature are reported. Using the Celsius scale as an example, an object's temperature is given as °C while a change in its temperature is given as C°.

12.2 THE KELVIN TEMPERATURE SCALE

- The symbol **K** is not written with the degree symbol, nor is the word "degrees" used when reporting Kelvin temperature.
- The **kelvin** is the SI unit for temperature.
- The size of one kelvin is identical to the size of one Celsius degree.
- The lowest possible temperature below which no substance can be cooled is the zero point on the Kelvin scale, and is referred to as **absolute zero**.
- The Kelvin temperature T and the Celsius temperature T_C are related by

$$T = T_C + 273.15 \qquad (12.1)$$

- When a gas is heated, its pressure increases, and when a gas is cooled, its volume decreases, assuming the gas is confined to a fixed volume. This is the basis for the **constant-volume gas thermometer**.
- If the pressure of a low-density-gas is plotted versus temperature, a straight line is obtained. If the straight line is extrapolated, it crosses the temperature axis at -273.15°C. This is the

absolute zero point for temperature measurement.

12.3 THERMOMETERS

- A property that changes with temperature is called a **thermometric property**.
- The **thermocouple** is a thermometer that consists of thin wires of different metals, welded together at the ends to form junctions. The thermocouple generates a "voltage" that depends on the *difference in temperature* between the two junctions.
- **Electric resistance thermometers** utilize the fact that most substances offer resistance to the flow of electricity that changes with temperature.
- The infrared radiation emitted by an object can be used to produce a kind of "thermal painting" of the body that is called a **thermograph or thermogram**.

12.4 LINEAR THERMAL EXPANSION

- **Linear Thermal Expansion of a Solid:**
  The length L_o of an object changes by an amount ΔL when its temperature changes by an amount ΔT:

  $$\Delta L = \alpha L_o \Delta T \qquad (12.2)$$

  where α is the **coefficient of linear expansion**.
- Common Unit for the Coefficient of Linear Expansion: $1/C° = (C°)^{-1}$
- The value of α depends on the nature of the material.
- A **bimetallic strip** is made from two thin strips of metal that have *different* coefficients of linear expansion.
- A hole in a piece of solid material expands when heated and contracts when cooled, just as if it were filled with the material that surrounds it.

12.5 VOLUME THERMAL EXPANSION

- The volume of a normal material increases as the temperature increases. Most solids and liquids behave in this fashion.
- **Volume Thermal Expansion**
  The volume V_o of an object changes by an amount ΔV when its temperature changes by an amount ΔT:

  $$\Delta V = \beta V_o \Delta T \qquad (12.3)$$

  where β is the **coefficient of volume expansion.**
- Common Unit for the Coefficient of Volume Expansion: $(C°)^{-1}$.
- For most solids, the coefficient of volume expansion is three times greater than the coefficient of linear expansion: $\beta = 3\alpha$.
- If a cavity exists within a solid object, the volume of the cavity expands when the object expands, just as if the cavity were filled with the surrounding material.
- If water at 0°C is heated its volume *decreases* until it reaches 4°C.

12.6 HEAT AND INTERNAL ENERGY

- **Heat** is energy that flows from a higher-temperature object to a lower-temperature object.
- SI Unit of Heat: joule (J).
- The heat that flows from hot to cold originates in the internal energy of the hot substance.
- The **internal energy** of a substance is the sum of the molecular kinetic energy, the molecular potential energy, and other kinds of molecular energy.
- The word "heat" is used only when referring to the energy actually in transit from hot to cold.

12.7 HEAT AND TEMPERATURE CHANGE: SPECIFIC HEAT CAPACITY

- The heat Q that must be supplied or removed to change the temperature of a substance of mass m by an amount ΔT is

$$Q = cm\Delta T \qquad (12.4)$$

where c is the **specific heat capacity** of the substance.

- Common Unit for Specific Heat Capacity: J/(kg·C°).
- The specific heat capacity depends on the nature of the material.
- One **kilocalorie** (1 kcal) was defined historically as the amount of heat needed to raise the temperature of one kilogram of water by one Celsius degree.
- The specific heat capacity of water is 1.00 kcal/(kg·C°).
- One **calorie** (1 cal) is defined as the amount of heat needed to raise the temperature of one gram of water by one Celsius degree.
- One **Calorie** = 1000 calories = 1 kcal. This is the unit used by nutritionists.
- **One British thermal unit** (Btu) is the amount of heat needed to raise the temperature of one pound of water by one Fahrenheit degree.
- 1 kcal = 4186 joules or 1 cal = 4.186 joules is the conversion factor known as the **mechanical equivalent of heat**.
- Different values are obtained when the specific heat capacity for a gas is measured under

conditions of *constant pressure*, c_P, as compared to conditions of *constant volume*, c_V. The value c_P is greater than the value c_V.

- If there is no heat loss to the external surroundings, the heat lost by the hotter objects equals the heat gained by the cooler ones, a process that is consistent with the conservation of energy.
- A **calorimeter** is essentially an insulated container, and is used in a technique known as **calorimetry**.

12.8 HEAT AND PHASE CHANGE: LATENT HEAT

- There is more than one **phase** of matter.
- Matter can change from one phase to another, and heat plays a role in the change.
- A solid can **melt** or **fuse** into a liquid if heat is added, while the liquid can **freeze** into a solid if heat is removed.
- A liquid can **evaporate** into a gas if heat is supplied, while the gas can **condense** into a liquid if heat is taken away. Rapid evaporation, with the

- formation of bubbles within the liquid, is called **boiling**.
- A solid sometimes can **sublime** directly into a gas if heat is added. If heat is removed under the right conditions, the gas will condense directly into a solid.
- The **latent heat** L is the heat per kilogram that must be added or removed when a substance changes from one phase to another at a constant temperature.
- The **latent heat of fusion** L_f refers to the change between solid and liquid phases.
- The **latent heat of vaporization** L_v applies to the change between liquid and gas phases.
- The **latent heat of sublimation** L_s refers to the change between solid and gas phases.

12.9 EQUILIBRIUM BETWEEN PHASES OF MATTER

- Under specific conditions of temperature and pressure a substance can exist in equilibrium in more than one phase at the same time.

- The pressure of the vapor that coexists in equilibrium with the liquid is called the **equilibrium vapor pressure** of the liquid.
- The equilibrium vapor pressure does not depend on the volume of space above the liquid.
- The equilibrium vapor pressure depends only on the temperature of the liquid: a higher temperature causes a higher pressure.
- Only when the temperature and vapor pressure correspond to a point on a curved line, which is called the **vapor pressure curve** or **vaporization curve**, can liquid and vapor phases coexist in equilibrium.
- A liquid boils at the temperature for which the vapor pressure equals the external pressure.
- For solid/liquid equilibrium, a plot of the equilibrium pressure versus equilibrium temperature is referred to as the **fusion curve**.

12.10 HUMIDITY

- The **partial pressure** of a gas in a mixture of gases is the pressure it would exert if it alone occupied the entire volume at the same temperature as the mixture.

- Percent **relative humidity** =

  $$\frac{\text{Partial pressure of water vapor}}{\text{Equilibrium vapor pressure of water at the existing temperature}} \times 100 \quad (12.5)$$

- When the relative humidity is 100%, the water vapor is said to be **saturated**.
- When the relative humidity is less than 100%, the water vapor is said to be **unsaturated**.
- The temperature known as the **dew point** is the temperature at which the partial pressure of the vapor equals the equilibrium vapor pressure.

CHAPTER 13 | ***THE TRANSFER OF HEAT***

This chapter introduces the concept of heat transfer and the mechanisms by which it occurs. Convection, conduction, and radiation are all covered.

With respect to conduction, the related concepts of insulators and conductors are considered, along with the properties that make particular substances good heat conductors or insulators.

The important electromagnetic waves are defined so that radiation can properly be presented.

Important Concepts

- convection current
- convection
- thermal
- inversion layer
- natural convection
- forced convection
- conduction
- metals
- thermal conductors
- thermal insulators
- thermal conductivity
- electromagnetic waves
- radiation

- perfect blackbody
- blackbody
- Stefan-Boltzmann constant
- emissivity e
- radiant power
- R value

13.1 CONVECTION

- When heat is transferred to or from a substance, the internal energy of the substance can change.
- Virtually all our energy originates in the sun and is transferred to us through the void of space.
- When part of a fluid is warmed the volume of the fluid expands, and the density decreases. The surrounding cooler and denser fluid exerts a buoyant force on the warmer fluid, and pushes it upward. The fluid flow carries along heat and is called a **convection current**.
- **Convection** is the process in which heat is carried from place to place by the bulk movement of a fluid.

- A **thermal** occurs when the warmed ground heats the surrounding air, which is pushed upward by the surrounding cooled and denser air.
- It is usual for temperature to decrease with increasing altitude.
- If the temperature increases with increasing altitude in a layer of the atmosphere, the layer is called an **inversion layer**.
- **Forced convection** uses an external device such as a pump.

13.2 CONDUCTION

- **Conduction** is the process whereby heat is transferred directly through a material, any bulk motion of the material playing no role in the transfer.
- One mechanism for conduction occurs when the atoms or molecules in a hotter part of the material vibrate or move with greater energy than those in a cooler part. By means of collisions, the more energetic molecules pass on some of their energy to their less energetic neighbors.

- Metals are different from most substances in having a pool of electrons that are more or less free to wander throughout the metal. These free electrons can transport energy and allow metals to transfer heat very well.
- Those materials that conduct heat well are called **thermal conductors**, while those that conduct heat poorly are known as **thermal insulators**.
- The amount of heat Q conducted through a bar from the warmer end to the cooler end depends on a number of factors:
    1. Q is proportional to the length of time t during which conduction takes place.
    2. Q is proportional to the temperature difference ΔT between the two ends of the bar.
    3. Q is proportional to the cross-sectional area of the bar.
    4. Q is inversely proportional to the length L of the bar.
- The heat Q conducted during a time t through a bar of length L and cross-sectional area A is

$$Q = \frac{(kA\Delta T)t}{L} \qquad (13.1)$$

where ΔT is the temperature difference between the ends of the bar and k is the **thermal conductivity** of the material.

- SI Unit of Thermal Conductivity: $J/(s \cdot m \cdot C°)$.
- The thermal conductivity is also given in units of $W/(m \cdot C°)$.
  Different materials have different thermal conductivities. Metals are good thermal conductors. Liquids and gases generally have small thermal conductivities.
- In most fluids the heat transferred by conduction is negligible compared to that transferred by convection when there are strong convection currents.

13.3 RADIATION

- Energy from the sun is brought to earth by a class of waves known as **electromagnetic waves**.
- **Radiation** is the process in which energy is transferred by means of electromagnetic waves.
- The term **perfect blackbody** or simply **blackbody** is used when referring to an object that absorbs *all* the electromagnetic waves falling on it.

- A material that is a good absorber is also a good emitter, and a material that is a poor absorber is also a poor emitter.
- The **Stefan-Boltzmann constant** is the proportionality constant that relates Q to T^4 for a perfect blackbody. It is $\sigma = 5.67 \times 10^{-8} \text{J}/(\text{s}\cdot\text{m}^2\cdot\text{K}^4)$.
- The **emissivity** e is the ratio of the energy an object actually radiates to the energy the object would radiate if it were a erfect emitter.
- The emissivity is a dimensionless number between zero and one.
- **The Stefan-Boltzmann Law of Radiation**
  The **radiant energy** Q, emitted in a time t by an object that has a Kelvin temperature T, a surface area A, and an emissivity e, is given by

$$Q = e\sigma T^4 At \qquad (13.2)$$

where σ is the Stefan-Boltzmann constant and has the value of $5.67 \times 10^{-8} \text{J}/(\text{s}\cdot\text{m}^2\cdot\text{K}^4)$. An object emits a net **radiant power** $P_{net} = (Q/t)_{net}$. The net power is the power the object emits minus the power it absorbs. The expression for P_{net} when the temperature of the object is T and the temperature of the environment is T_o is

$$P_{net} = e\sigma A(T^4 - T_o^4) \qquad (13.3)$$

13.4 APPLICATIONS

- For an insulator, the heat per unit time Q/t flowing through a thickness of material is:

$$\frac{Q}{t} = \frac{A\Delta T}{L/k}$$

- The term L/k in the denominator is called the **R value** of the insulation.
- Larger R values mean better insulation.

CHAPTER 14 | ***THE IDEAL GAS LAW AND KINETIC THEORY***

This chapter takes the quantity of matter, or the number of molecules, into account when considering the behavior of gases. The ideal gas law, Boyle's law, and Charles' law are presented to model this behavior.

The kinetic energy of the gas molecules is examined, where each molecule is considered to have its own speed. The kinetic energies of molecules cause Brownian motion. The process of diffusion of a gas is also studied.

Important Concepts

- atomic mass
- atomic mass unit (u)
- molecular mass
- gram-mole
- Avogadro's number N_A
- ideal gas
- universal gas constant
- ideal gas law
- Boltzmann's constant
- standard temperature and pressure (STP)
- Boyle's law

- isotherm
- Charles' law
- root-mean-square speed
- Brownian motion
- monatomic ideal gas
- diffusion
- solvent
- solute
- Fick's law of diffusion

14.1 MOLECULAR MASS, THE MOLE, AND AVOGADRO'S NUMBER

- The relative masses of atoms of different elements can be expressed in terms of their **atomic masses**, which indicate how massive one atom is compared to another. The atomic masses are listed in the periodic table.
- The atomic mass scale is set up with a unit called the **atomic mass unit** (u). The reference element has been chosen to be the most common isotope of carbon, called carbon-12, and its atomic mass is defined to be exactly twelve atomic mass units.

- The **molecular mass** of a molecule is the sum of the atomic masses of its atoms.
- One **gram-mole** or simply one **mole** (mol) of a substance contains as many particles as there are atoms in 12 grams of the isotope carbon-12. This number is called **Avogadro's number** N_A, and is 6.022×10^{23}.
- The mole is the SI base unit for expressing the amount of a substance.
- One mole of a substance has a mass in grams that is equal to the atomic or molecular mass of the substance.

14.2 THE IDEAL GAS LAW

- An **ideal gas** is an idealized model for real gases that have sufficiently low densities. The condition of low density means that the molecules of the gas are so far apart that they do not interact (except during collisions that are effectively elastic).
- **Ideal Gas Law**
  The absolute pressure P of an ideal gas is directly proportional to the Kelvin temperature T and the number of moles n of the gas and is

inversely proportional to the volume V of the gas: $P = R(nT/V)$. In other words,

$$PV = nRT \qquad (14.1)$$

where R is the **universal gas constant** and has the value of 8.31 J/(mol·K) in SI units.

- Sometimes it is convenient to express the ideal gas law in terms of the total number of particles N, instead of the number of moles n.

$$PV = N\left(\frac{R}{N_A}\right)T$$

- The constant term R/N_A is referred to as **Boltzmann's constant** and is represented by the symbol k:

$$k = \frac{R}{N_A} = 1.38 \times 10^{-23} \text{ J/K}$$

- The ideal gas law becomes

$$PV = NkT \qquad (14.2)$$

- A temperature of 273 K and a pressure of one atmosphere (1.013×10^5 Pa) are known as **standard temperature and pressure** (STP).

- For an ideal gas where T and n are constant, the ideal gas follows **Boyle's law**:

$$P_i V_i = P_f V_f \qquad (14.3)$$

- When pressure and volume are plotted for constant-temperature conditions, the resulting curve is called an **isotherm**, meaning "same temperature."
- For an ideal gas where P and n are constant, the ideal gas follows **Charles' law**:

$$\frac{V_i}{T_i} = \frac{V_f}{T_f} \qquad (14.4)$$

14.3 KINETIC THEORY OF GASES

- Gas particles are in constant, random motion, colliding with each other and with the walls of the container.
- For a gas, the atoms or molecules, in general, have different speeds. It is possible, however, to speak about an average particle speed.
- For gas atom or molecule colliding elastically and perpendicularly with its container walls at

velocity v, Newton's second law of motion, in the form of the impulse-momentum theorem, gives the average force exerted on the particle by the wall as:

$$\text{Average force} = \frac{(-mv) - (+mv)}{2L/v} = \frac{-mv^2}{L}$$

where L is the width of the container.

- Since N particles move randomly in three dimensions, one third of them on the average strike the right wall during time t. Therefore, the total magnitude of the force is

$$F = \left(\frac{N}{3}\right)\left(\frac{\overline{mv^2}}{L}\right)$$

In this result v^2 has been replaced by $\overline{v^2}$, the average value of the square speed. The collision of particles possesses a Maxwell distribution of speeds, so the average value for v^2 must be used.

- The square root of the quantity $\overline{v^2}$ is called the **root-mean-square speed**, or, for short, the *rms speed*; $v_{rms} = \sqrt{\overline{v^2}}$. Using this definition, it can be shown that

$$PV = \frac{2}{3} N (\frac{1}{2} m v_{rms}^2) \qquad (14.5)$$

- The term $\frac{1}{2} m v_{rms}^2$ is the average translational kinetic energy $\overline{KE}$ of an individual particle.
- Therefore,

$$\overline{KE} = \frac{1}{2} m v_{rms}^2 = \frac{3}{2} kT \qquad (14.6)$$

- On the average, gas particles have greater kinetic energies when the gas is hotter than when it is cooler.
- If two ideal gases have the same temperature, the relation $\frac{1}{2} m v_{rms}^2 = \frac{3}{2} kT$ indicates that the average kinetic energies of each kind of gas particle is the same. In general, however, the rms speeds of the different particles are not the same, for the masses may be different.
- **Brownian motion** can be explained as a response of large suspended particles moving in response to impacts from the moving molecules of the fluid medium. As a result, the suspended particles have the same average translational kinetic energy as the fluid molecules.

- A **monatomic ideal gas** is composed of single atoms. These atoms are so small that the mass is concentrated almost at a point.
- The internal energy for a monatomic ideal gas is

$$U = \frac{3}{2}nRT \qquad (14.7)$$

14.4 DIFFUSION

- The process in which molecules move from a region of higher concentration to one of lower concentration is called **diffusion**.
- Compared to the rate of diffusion in gases, the rate is generally smaller in liquids and even smaller in solids.
- The host medium for diffusion is referred to as the **solvent**, while the diffusing substance is known as the **solute**.
- Relatively speaking, diffusion is a slow process, even in gases.
- **Fick's Law of Diffusion**
  The mass m of solute that diffuses in a time t through a solvent contained in a channel of length L and cross-sectional area A is

$$m = \frac{(DA\Delta C)t}{L} \qquad (14.8)$$

where ΔC is the concentration difference between the ends of the channel and D is the diffusion constant.
- SI Unit for Diffusion Constant: m^2/s

CHAPTER 15 | *THERMODYNAMICS*

This chapter presents the laws of thermodynamics, and introduces the concept of entropy. Adiabatic, isothermal, isochoric, and isobaric processes are looked at in detail.

The concepts of thermodynamics are explored for practical applications, such as refrigeration and the Carnot engine. These applications make use of heat reservoirs and of specific heat capacities.

Important Concepts

- thermodynamics
- diathermal wall
- adiabatic wall
- thermal equilibrium
- zeroth law of thermodynamics
- first law of thermodynamics
- quasi-static process
- isobaric process
- isochoric process
- isothermal process
- adiabatic process
- molar specific heat capacity
- second law of thermodynamics

- efficiency of an engine
- reversible process
- Carnot's principle
- thermodynamic temperature scale
- refrigeration process
- heat pump
- entropy
- third law of thermodynamics

15.1 THERMODYNAMIC SYSTEMS AND THEIR SURROUNDINGS

- **Thermodynamics** is the branch of physics that is built upon the fundamental laws that heat and work obey.
- In thermodynamics the collection of objects upon which attention is being focused is called the **system**, while everything else in the environment is called the **surroundings**.
- Walls that permit heat to flow through them are called **diathermal walls**.
- Perfectly insulating walls that do not permit heat to flow between the system and its surroundings are called **adiabatic walls**.

- The physical condition of a system is the **state of a system**.

15.2 THE ZEROTH LAW OF THERMODYNAMICS

- Two systems are said to be in **thermal equilibrium** if there is no net flow of heat between them when they are brought into thermal contact.
- Temperature is the indicator of thermal equilibrium in the sense that there is no net flow of heat between two systems in thermal contact that have the same temperature.
- **Zeroth Law of Thermodynamics**
  Two systems individually in thermal equilibrium with a third system are in thermal equilibrium with each other.
- The zeroth law implies that all parts of a system must be in thermal equilibrium if the system is to have a definable single temperature.

15.3 THE FIRST LAW OF THERMODYNAMICS

- **The First Law of Thermodynamics**
  The internal energy of a system changes from an initial value U_i to a final value of U_f due to heat Q and work W:

  $$\Delta U = U_f - U_i = Q - W \qquad (15.1)$$

  Q is positive when the system gains heat and negative when it loses heat. W is positive when work is done by the system and negative when work is done on the system.
- The value of ΔU is determined for an ideal gas once the initial and final temperatures are specified, because the internal energy of an ideal gas depends only on the Kelvin temperature.
- The internal energy depends only on the state of a system, not the method by which the system arrives at a given state. Internal energy, therefore, is referred to as a *function of state*.
- Heat and work are not functions of state, because they have different values for each

different method used to make the system change from one state to another.

15.4 THERMAL PROCESSES

- A process in this section is assumed to be **quasi-static**, which means that it occurs slowly enough that a uniform pressure and temperature exist throughout all regions of the system at all times.
- An **isobaric process** is one that occurs at constant pressure.
- The expression for work for an isobaric process is

$$W = P\Delta V = P(V_f - V_i) \qquad (15.2)$$

This result predicts a positive value for work done by a system when a system expands adiabatically. For isobaric compression the work is negative.
- An **isochoric process** is one that occurs at a constant volume.
- Since no work is done by an isochoric process, the first law of thermodynamics indicates that

the heat serves only to change the internal energy of the system.
- An **isothermal process** is one that takes place at constant temperature.
- An **adiabatic process** is one that occurs without the transfer of heat.
- When work is done by a system adiabatically, the internal energy of the system decreases by exactly the amount of the work done.
- When work is done on a system adiabatically, the internal energy increases correspondingly.
- The area under a pressure-volume graph is the work for any kind of process.

15.5 THERMAL PROCESSES THAT UTILIZE AN IDEAL GAS

- When a system performs work isothermally, the temperature remains constant.
- The work associated with the isothermal expansion or compression of an ideal gas is

$$W = nRT \ln\left(\frac{V_f}{V_i}\right) \quad (15.3)$$

- Since internal energy of an ideal gas is proportional to the Kelvin temperature, the internal energy remains constant throughout an isothermal process.
- The work associated with an adiabatic expansion or compression of a monatomic ideal gas is

$$W = \frac{3}{2}nR(T_i - T_f) \quad (15.4)$$

- When an ideal gas expands adiabatically, it does positive work. Therefore, the term $T_i - T_f$ is positive, and the final temperature of the gas must be less than the initial temperature.
- The reverse of an adiabatic expansion is an adiabatic compression, for which the final temperature of an ideal gas must exceed the initial temperature.
- The equation that gives the adiabatic curve between the initial pressure and volume (P_i, V_i) and the final pressure and volume (P_f, V_f) is

$$P_i V_i^{\gamma} = P_f V_f^{\gamma} \quad (15.5)$$

- where the exponent γ is the ratio c_P/c_V of the specific heat capacities at constant pressure and constant volume.

15.6 SPECIFIC HEAT CAPACITIES AND THE FIRST LAW OF THERMODYNAMICS

- We replace the expression $Q = cm\Delta T$ with the following analogous expression:

$$Q = Cn\Delta T \qquad (15.6)$$

where the capital C refers to the **molar specific heat capacity** in units of J/(mol·K).

- For gases it is necessary to distinguish between the molar specific heat capacities C_P and C_V, which apply, respectively, to conditions of constant pressure and constant volume.
- The molar specific heat capacities for a monatomic ideal gas at constant pressure and constant volume, respectively, are:

$$C_P = \frac{3}{2}R + R = \frac{5}{2}R \qquad (15.7)$$

$$C_V = \frac{3}{2}R \qquad (15.8)$$

- The ratio γ of the specific heats is

$$\gamma = \frac{C_P}{C_V} = \frac{\frac{5}{2}R}{\frac{3}{2}R} = \frac{5}{3} \qquad (15.9)$$

- For real monatomic gases near room temperature the experimental values of γ are very close to the theoretical value of 5/3.
- The difference between C_P and C_V arises because work is done when the gas expands in response to the addition of heat under constant pressure, whereas no work is done under conditions of constant volume. For an ideal gas, C_P exceeds C_V by an amount equal to R, the ideal gas constant:

- $$C_P - C_V = R \qquad (15.10)$$

15.7 THE SECOND LAW OF THERMODYNAMICS

- **The Second Law of Thermodynamics: The Heat Flow Statement**

- Heat flows spontaneously from a substance at a higher temperature to a substance at a lower temperature and does not flow spontaneously in the reverse direction.

15.8 HEAT ENGINES

- A **heat engine** is any device that uses heat to perform work.
- The subscript H stands for "hot," and the place from which the input heat comes is called the "**hot reservoir**." The subscript C means "cold," and the place in the environment where the rejected heat goes is known as the "**cold reservoir**." The symbols Q_H, Q_C, and W refer to magnitudes only, without reference to algebraic sign.
- The **efficiency** e of a heat engine is defined as the ratio of the work W done by the engine to the input heat Q_H:

$$e = \frac{\text{Work done}}{\text{Input heat}} = \frac{W}{Q_H} \qquad (15.11)$$

- Efficiencies are often quoted as percentages obtained by multiplying the ratio W/Q_H by a factor of 100.
- Some of the engine's input heat Q_H is converted to work W, and the remainder Q_C is rejected to the cold reservoir. If there are no losses in the engine, the principle of energy conservation requires that

$$Q_H = W + Q_C \qquad (15.12)$$

- An alternative expression for the efficiency e of a heat engine is

$$e = \frac{Q_H - Q_C}{Q_H} = 1 - \frac{Q_C}{Q_H} \qquad (15.13)$$

15.9 CARNOT'S PRINCIPLE AND THE CARNOT ENGINE

- A reversible process is one in which both the system and its environment can be returned to exactly the states they were in before the process occurred.

- A process that involves an energy-dissipating mechanism, such as friction, cannot be reversible.
- **Carnot's Principle: An Alternative Statement of the Second Law of Thermodynamics**
  No irreversible engine operating between two reservoirs at constant temperatures can have a greater efficiency than a reversible engine operating between the same temperatures. Furthermore, all reversible engines operating between the same temperatures have the same efficiency.
- Carnot's principle does *not* state, or even imply, that a reversible engine has an efficiency of 100%.
- If Carnot's principle were not valid, it would be possible for heat to flow spontaneously from a cold substance to a hot substance.
- No real engine operates reversibly.
- The idea of a reversible engine provides a useful standard for judging the performance of real engines.
- An important feature of the Carnot engine is that all input heat Q_H originates from a hot reservoir at a single temperature T_H and all rejected heat Q_C goes into a cold reservoir at a single temperature T_C.

- The **thermodynamic temperature scale** is defined such that the ratio of the temperatures of the cold and hot reservoirs is equal to Q_C / Q_H.

$$Q_C/Q_H = T_C/T_H \qquad (15.14)$$

where the temperatures T_C and T_H must be expressed in kelvins.

- Efficiency of a Carnot engine $= 1 - \dfrac{T_C}{T_H}$ (15.15)

- The efficiency of a Carnot engine is the maximum possible efficiency for a heat engine operating between two Kelvin temperatures T_C and T_H.

- Even a perfect heat engine has an efficiency that is less than 1.0 or 100%.

- It is not possible to cool a substance to absolute zero, so nature does not permit the existence of a 100% efficient heat engine.

15.10 REFRIGERATORS, AIR CONDITIONERS, AND HEAT PUMPS

- If work is used, heat can be made to flow from cold to hot, against its natural tendency.
- The use of work W to extract an amount of heat Q_C from a cold reservoir and deposit an amount of heat Q_H in a hot reservoir is called a **refrigeration process**.
- Energy is conserved during a refrigeration process, so $Q_H = W + Q_C$.
- If the process occurs reversibly, we have ideal devices that are called Carnot refrigerators, Carnot air conditioners, and Carnot heat pumps.
- In a **refrigerator**, the interior of the unit is the cold reservoir, while the warmer exterior is the hot reservoir.
- In an **air conditioner** the room itself is the cold reservoir and the outdoors is the hot reservoir.
- The coefficient of performance for a refrigerator or an air conditioner
$$= Q_C/W. \qquad (15.16)$$
- The **heat pump** uses work to make heat Q_C from the wintry outdoors flow into the warm house.

- The coefficient of performance for a heat pump = Q_H/W. (15.17)

15.11 ENTROPY AND THE SECOND LAW OF THERMODYNAMICS

- The loss of the ability to perform work in an irreversible process can be expressed in terms of a concept called **entropy**.

- The quantity $(Q/T)_R$ is called the change in entropy ΔS:

$$\Delta S = (Q/T)_R \qquad (15.18)$$

In this expression the temperature T must be in kelvins, and the subscript R refers to the word "reversible."
- Entropy is a function of the state or condition of the system.
- Reversible processes do not alter the total entropy of the universe.
- Any irreversible process increases the entropy of the universe.
- **The Second Law of Thermodynamics Stated in Terms of Entropy**
  The total entropy of the universe does not change when a reversible process occurs

($\Delta S_{universe} = 0$) and increases when an irreversible process occurs ($\Delta S_{universe} > 0$).
- Since irreversible processes cause the entropy of the universe to increase, they cause energy to be degraded, for part of the energy becomes unavailable for the performance of work according to

$$W_{unavailable} = T_o \Delta S_{universe} \qquad (15.19)$$

- where T_o is the Kelvin temperature of the coldest heat reservoir.
  An increase in entropy is associated with an increase of disorder.

15.12 THE THIRD LAW OF THERMODYNAMICS

- **The Third Law of Thermodynamics**
  It is not possible to lower the temperature of any system to absolute zero in a finite number of steps.

CHAPTER 16 | *WAVES AND SOUND*

This chapter presents a discussion of both transverse and longitudinal waves, including such properties as wavelength, frequency, amplitude, period, pitch, and the speed of the disturbance. The waves considered are all periodic.

An understanding of energy is necessary for this chapter.

A significant amount of this chapter is devoted to sound, including the property of loudness. Material from previous chapters regarding gases and liquids are incorporated into the presentation.

Important Concepts

- wave
- transverse wave
- longitudinal wave
- periodic wave
- cycle
- amplitude
- wavelength
- period
- frequency
- linear density
- sound

- condensation
- rarefaction
- pure tone
- infrasonic
- ultrasonic
- pitch
- pressure amplitude
- loudness
- adiabatic bulk modulus
- power
- sound intensity
- threshold of hearing
- decibel
- intensity level
- sonar
- Doppler effect
- Fletcher-Munson curves

16.1 THE NATURE OF WAVES

- A **wave** is a traveling disturbance.
- A wave carries energy from place to place.
- A **transverse wave** is one in which the disturbance is perpendicular to the direction of travel of the wave.

- A **longitudinal wave** is one in which the disturbance is parallel to the line of travel of the wave.
- A **sound wave** is a longitudinal wave.
- Some waves are neither transverse nor longitudinal.

16.2 PERIODIC WAVES

- **Periodic waves** consist of patterns that are produced over and over again by the source.
- Some of the terminology **(cycle, amplitude, period, and frequency)** used to describe periodic waves is the same as that given in Chapter 10 for simple harmonic motion.
- The **amplitude** A is the maximum excursion of a particle of the medium from the particle's undisturbed position.
- The **wavelength** λ is the horizontal length of one cycle of the wave.
- The **period** T is the time required for the wave to travel the distance of one wavelength.
- The period T is related to the **frequency** f:

$$f = \frac{1}{T} \qquad (10.9)$$

- The period is commonly measured in seconds, and the frequency is measured in cycles per second or **hertz** (Hz).
- A wave's **speed** v is related to it's wavelength (of frequency) and its period by:

$$v = \frac{\lambda}{T} = f\lambda \qquad (16.1)$$

16.3 THE SPEED OF A WAVE ON A STRING

- The ability of one particle to pull on its neighbors depends on how tightly the string is stretched, that is, on the tension. The greater the tension, the faster the wave travels.
- Other things being equal, a wave travels faster on a string whose particles have a small mass, or a small mass per unit length.
- The mass per unit length is called the **linear density**.
- The speed v of a small amplitude wave on a string is given by:

$$v = \sqrt{\frac{F}{m/L}} \qquad (16.2)$$

16.4 THE MATHEMATICAL DESCRIPTION OF A WAVE

- When a wave travels through a medium, it displaces the particles of the medium from their undisturbed positions.
- Equation 16.3 represents the displacement of a particle caused by a wave traveling in the +x direction (to the right) that has an amplitude A frequency f, and wavelength λ. Equation 16.4 applies to a wave moving in the -x direction (to the left).

$$y = A \sin\left(2\pi ft - \frac{2\pi x}{\lambda}\right) \qquad (16.3)$$

$$y = A \sin\left(2\pi ft + \frac{2\pi x}{\lambda}\right) \qquad (16.4)$$

These equations apply to transverse or longitudinal waves and assume that $y = 0$ when $x = 0$ and $t = 0$.
- $(2\pi ft - 2\pi x/\lambda)$ is called the **phase angle** for a wave traveling in the +x direction.

- A particle located at a distance x also exhibits simple harmonic motion, but its phase angle is

$$2\pi ft - \frac{2\pi x}{\lambda} = 2\pi f\left(t - \frac{x}{f\lambda}\right) = 2\pi f\left(t - \frac{x}{\upsilon}\right)$$

  The quantity x/υ is the time needed for the wave to travel the distance x.

- *When using a calculator to evaluate the function sin ($2\pi ft - 2\pi x/\lambda$) the calculator must be set to its radian mode.*

16.5 THE NATURE OF SOUND

- **Sound** is a longitudinal wave that is caused by a vibrating object.
- Sound can be created only in a material medium.
- The region of increased pressure is called a **condensation** and travels away from a source.
- A region known as a **rarefaction** occurs where the air pressure is slightly less than normal.
- The **wavelength** λ is the distance between centers of two successive condensations; λ is

also the distance between the centers of two successive rarefactions.
- Each cycle of a sound wave includes one condensation and one rarefaction, and the **frequency** is the number of cycles per second that pass a given location.
- A **pure tone** is sound with a single frequency.
- Sound waves with frequencies below 20 Hz are said to be **infrasonic**, while those with frequencies above 20 kHz are referred to as **ultrasonic**.
- The **pressure amplitude** of a wave is the magnitude of the maximum change in pressure, measured relative to the undisturbed pressure.
- **Loudness** is an attribute that depends primarily on the amplitude of the wave: the larger the amplitude, the louder the sound.

16.6 THE SPEED OF SOUND

- Near room temperature, the speed of sound in air is 343 m/s and is markedly greater in liquids and solids.
- In an ideal gas, the speed of sound is given by

$$v = \sqrt{\gamma\, kT/m} \qquad (16.5)$$

- The compression and expansion process is adiabatic.
- In liquids the speed of sound depends on the density ρ and the **adiabatic bulk modulus** B_{ad} of the liquid:

$$\upsilon = \sqrt{B_{ad}/\rho} \qquad (16.6)$$

The adiabatic bulk modulus is needed because the condensations and rarefactions occur under adiabatic rather than isothermal conditions.
- When sound travels through a long slender solid bar, the speed of the sound depends on the properties of the medium according to

$$\upsilon = \sqrt{Y/\rho} \qquad (16.7)$$

where Y is Young's modulus and ρ is the density.

16.7 SOUND INTENSITY

- Sound waves carry energy that can be used to do work.

- The amount of energy transported per second by a sound wave is called the **power** of the wave and is measured in SI units of joules per second (J/s) or watts (W).
- The **sound intensity** I is defined as the sound power P that passes perpendicularly through a surface divided by the area A of that surface:

$$I = \frac{P}{A} \qquad (16.8)$$

- The unit of sound is power per unit area, or W/m^2.
- For a 1000-Hz tone, the smallest sound intensity that the human ear can detect is about 1×10^{-12} W/m^2; this intensity is called the **threshold of hearing**.
- For spherically uniform radiation, the radiated power P passes through the spherical surface of area $A = 4\pi r^2$, so that the intensity at a distance r is

$$I = \frac{P}{4\pi r^2} \qquad (16.9)$$

16.8 DECIBELS

- The **decibel** (dB) is a measurement unit used when comparing two sound intensities.
- The **intensity level** β expressed in decibels is defined as follows:

$$\beta = 10 \log \left(\frac{I}{I_o} \right) \qquad (16.10)$$

  where "log" denotes the logarithm to the base ten and I_o is the intensity of the reference level to which I is being compared.
- Although β is called the "intensity level'" it is not an intensity.
- The decibel is unitless.
- An intensity level of zero decibels does not mean that the sound intensity I is zero; it means that $I = I_o$.
- Hearing tests have revealed that a one-decibel change in the intensity level is approximately the smallest change in loudness that an average listener can detect.
- If the intensity level increases by 10 dB, the new sound seems approximately twice as loud as the original sound.

16.9 APPLICATIONS OF SOUND

- Sonar is a technique for determining water depth and locating underwater objects.
- Ultrasound waves are used in medicine for diagnostic purposes.

16.10 THE DOPPLER EFFECT

- The **Doppler effect** is the change in pitch or frequency of the sound detected by an observer because the sound source and the observer have different velocities with respect to the medium of sound propagation.
- For a source moving toward a stationary observer

$$f' = f\left(\frac{1}{1 - \frac{v_s}{v}}\right) \qquad (16.11)$$

where f' is the frequency perceived by the observer.

- For a source moving away from a stationary observer

$$f' = f\left(\frac{1}{1+\dfrac{v_s}{v}}\right) \qquad (16.12)$$

where f' is the frequency perceived by the observer.

- For an observer moving toward a stationary source

$$f' = f\left(1+\frac{v_o}{v}\right) \qquad (16.13)$$

- For an observer moving away from a stationary source

$$f' = f\left(1-\frac{v_o}{v}\right) \qquad (16.14)$$

- For both the observer and the source moving

$$f' = f \left(\frac{1 \pm \dfrac{v_o}{v}}{1 \mp \dfrac{v_s}{v}} \right) \quad (16.15)$$

In the numerator, the plus sign applies when the observer moves toward the source, and the minus sign applies when the observer moves away from the source. In the denominator, the minus sign is used when the source moves toward the observer, and the plus sign is used when the source moves away from the observer.

16.11 THE SENSITIVITY OF THE HUMAN EAR

- The ear is not equally sensitive to all frequencies. The **Fletcher-Munson curves** show the intensity levels at which different frequencies become audible.

CHAPTER 17 | *THE PRINCIPLE OF LINEAR SUPERPOSITION AND INTERFERENCE PHENOMENA*

This chapter considers the consequences of several waves occurring together, or passing through each other. The mechanisms of constructive and destructive interference are handled

The chapter also investigates the phenomenon of standing waves.

Important Concepts

- principle of linear superposition
- constructive interference
- destructive interference
- coherent source
- diffraction
- beat
- beat frequency
- transverse standing wave pattern
- node
- antinode
- harmonic
- fundamental frequency

- overtone
- natural frequency
- complex sound wave

17.1 THE PRINCIPLE OF LINEAR SUPERPOSITION

- **The Principle of Linear Superposition**
  When two or more waves are present simultaneously at the same place, the resultant wave is the sum of the individual waves.
- The principle of linear superposition can be applied to all types of waves.

17.2 CONSTRUCTIVE AND DESTRUCTIVE INTERFERENCE OF SOUND WAVES

- When two waves meet condensation-to-condensation and rarefaction-to-rarefaction, they are said to be exactly in phase and to exhibit **constructive interference**.

- When two waves always meet condensation to rarefaction, they are said to be exactly out of phase and to exhibit **destructive interference.**
- Sources that produce wave patterns that do not shift relative to one another as time passes are called **coherent sources**.
- For two wave sources vibrating in phase, a difference in path lengths that is an integer number (1, 2, 3,...) of wavelengths leads to constructive interference; a difference in path lengths that is a half-integer number ($\frac{1}{2}, 1\frac{1}{2}, 2\frac{1}{2},...$) of wavelengths leads to destructive interference.
- One of the interesting consequences of interference is that the energy is redistributed.

17.3 DIFFRACTION

- The bending of a wave around an obstacle or the edges of an opening is called **diffraction**.
- All kinds of waves exhibit diffraction.
- The angle θ defines the first minimum intensity point on either side of the center in terms of the wavelength λ and slit width D.

$$\sin\theta = \frac{\lambda}{D} \qquad (17.1)$$

- The first minimum for a circular opening is defined by:

$$\sin\theta = 1.22\frac{\lambda}{D} \qquad (17.2)$$

- High frequency sound has a "narrow dispersion."
- Low frequency sound has a "wide dispersion."
- Loudspeaker designers utilize a large value of λ/D in the type of speaker known as a diffraction horn. If diffraction is to be an effective spreading mechanism, the width D must be parallel to the floor.

17.4 BEATS

- Periodic variations in loudness are called **beats** and result from the interference between two sound waves with slightly different frequencies.
- The number of times per second that the loudness rises and falls is the **beat frequency**

and is the difference between the two sound frequencies.

17.5 TRANSVERSE STANDING WAVES

- When a standing wave occurs with transverse waves it is known as a **transverse standing wave pattern**.
- The **nodes** are places that do not vibrate at all, and the **antinodes** are places where maximum vibration occurs.
- The frequencies in the series (f_1, $2f_1$, $3f_1$, etc.) are called **harmonics**. The first frequency in this series is referred to as the **fundamental frequency**. The other frequencies in this series are known as **overtones**.
- Standing waves arise because identical waves travel on a string in opposite directions and combine in accord with the principle of linear superposition.
- The frequency f_1 at which resonance occurs is sometimes called a **natural frequency**.
- A string can have a series of natural frequencies.

- The series of natural frequencies that lead to standing waves on a string fixed at both ends is given by

$$f_n = n\left(\frac{v}{2L}\right) \quad n = 1, 2, 3, 4,... \quad (17.3)$$

17.6 LONGITUDINAL STANDING WAVES

- Standing waves can also be formed from longitudinal waves.
- As in the transverse standing wave, the distance between two successive antinodes is one-half of a wavelength.
- For a tube open at both ends

$$f_n = n\left(\frac{v}{2L}\right) \quad n = 1, 2, 3, 4,... \quad (17.4)$$

- For a tube that is open at one end

$$f_n = n\left(\frac{v}{4L}\right) \quad n = 1, 3, 5,... \quad (17.5)$$

17.7 COMPLEX SOUND WAVES

- A **complex sound wave** consists of a mixture of the fundamental and harmonic frequencies.

CHAPTER 18 | **ELECTRIC FORCES AND ELECTRIC FIELDS**

This chapter investigates static electricity, starting with the parts of an atom. In particular, the Coulomb force that attracts or repels charges is discussed, along with the electric field associated with a static electric charge.

The concept of electrical induction is a central feature of this chapter.

A complete understanding of this material requires knowledge of forces.

Important Concepts

- electric charge
- coulomb
- electrically neutral
- law of conservation of electric charge
- electric force (electrostatic force)
- electrical conductor
- electrical insulator
- valence electron
- "free" electron
- charging by contact
- charging by induction

- Coulomb's law
- permittivity of free space
- test charge
- electric field
- parallel plate capacitor
- charge density
- electric field lines
- electric dipole
- shielding
- charge distribution
- Gaussian surface
- electric flux
- Gauss' law

18.1 THE ORIGIN OF ELECTRICITY

- An atom consists of a small, relatively massive nucleus that contains particles called protons and neutrons. Surrounding the nucleus is a diffuse cloud of orbiting electrons.
- **Electric charge** is an intrinsic property of protons and electrons, and only two types of charge have been discovered, positive and negative. A proton has a positive charge, and an electron has a negative charge. A neutron has no electric charge.

- The magnitude of the charge on the proton exactly equals the magnitude of the charge on the electron; the proton carries a charge $+e$, and the electron carries a charge $-e$.
- The SI unit for measuring the magnitude of an electric charge is the **coulomb** (C).

$$e = 1.60 \times 10^{-19} \text{ C}$$

It should be noted that the symbol e represents only the magnitude of the charge on a proton or an electron, and does not include the algebraic sign that indicates whether the charge is positive or negative.
- When an object carries no net charge, the object is said to be **electrically neutral**.
- The charge on an electron or a proton is the *smallest* amount of free charge that has been discovered. Any charge of magnitude q is an integer multiple of e; that is, $q = Ne$, where N is an integer. Therefore, electric charge is said to be quantized.

18.2 CHARGED OBJECTS AND THE ELECTRIC FORCE

- Usually electrons are transferred, and the body that gains electrons acquires an excess of negative charge. The body that loses electrons has an excess of positive charge.
- **Law of Conservation of Electric Charge** During any process, the net electric charge of an isolated system remains constant (is conserved).
- Two electrically charged objects exert a force on one another.
- **Like charges repel and unlike charges attract each other.**
- The force on charged objects is an **electric force** (also called an electrostatic force).

18.3 CONDUCTORS AND INSULATORS

- Not only can electric charge exist on an object, but it can also move through an object.

- Materials differ vastly in their abilities to allow electric charge to move or be conducted through them.
- Substances that readily conduct electric charge are called **electrical conductors**.
- Most good thermal conductors are also good electrical conductors.
- Materials that conduct electric charge poorly are known as **electrical insulators**.
- Most thermal insulators are also electrical insulators.
- The outermost electrons in an atom are called **valence electrons**.
- In a good conductor, some valence electrons actually become detached from a parent atom and wander more or less freely throughout the material, belonging to no one atom in particular. The exact number of electrons detached from each atom depends on the nature of the material, but is usually between one and three.
- In an insulator virtually every electron remains bound to its parent atom.

18.4 CHARGING BY CONTACT AND BY INDUCTION

- The process of giving one object a net electric charge by placing it in contact with a charged object is known as **charging by contact**.
- Under most conditions the earth is a good electrical conductor.
- The process of giving one object a net electric charge without touching the object to a second charged object is called **charging by induction**.

18.5 COULOMB'S LAW

- **Coulomb's Law**
  The magnitude F of the electrostatic force exerted by one point charge on another point charge is directly proportional to the magnitudes q_1 and q_2 of the charges and inversely proportional to the square of the distance r between them:

$$F = k\frac{q_1 q_2}{r^2} \qquad (18.1)$$

where k is a proportionality constant whose value in SI units is $k = 8.99 \times 10^9$ N·m^2/C^2. The electrostatic force is directed along the line joining the charges, and it is attractive if the charges have unlike signs and repulsive if the charges have like signs.

- It is common practice to express k in terms of another constant ε_o, by writing $k = 1/(4\pi\varepsilon_o)$; ε_o is called the **permittivity of free space** and has the value of
$\varepsilon_o = 1/(4\pi k) = 8.85 \times 10^{-12}$ C^2/(N·m^2).

18.6 THE ELECTRIC FIELD

- It is useful to think of q_o as a **test charge** for determining the extent to which the surrounding charges generate a force. A test charge must be one with a very small magnitude, so it does not alter the location of the other charges.
- **Definition of an Electric Field**

The electric field **E** that exists at a point is the electrostatic force experienced by a small test charge q_o placed at that point divided by the charge itself:

$$\mathbf{E} = \frac{\mathbf{F}}{q_0} \qquad (18.2)$$

The electric field is a vector, and its direction is the same as the direction of the force **F** on a positive test charge.

- SI Unit of Electric Field: newton per coulomb (N/C)
- It is the surrounding charges that create an electric field at a given point.
- To determine the net field, it is necessary to obtain the various contributions separately and then find the vector sum of them.
- The electric field produced by a point charge has a magnitude:

$$E = \frac{kq}{r^2} \qquad (18.3)$$

- A **parallel plate capacitor** consists of two parallel metal plates, each with an area A. A charge $+q$ is spread uniformly over one plate, while a charge $-q$ is spread uniformly over the other plate.

- For a parallel plate capacitor

$$E = \frac{q}{\varepsilon_o A} = \frac{\sigma}{\varepsilon_o} \qquad (18.4)$$

In this expression the Greek symbol (σ) denotes the charge per unit area ($\sigma = q/A$) and is sometimes called the **charge density**. The field does not depend on the distance from the charges.

18.7 ELECTRIC FIELD LINES

- **Electric field lines** are sometimes called lines of force.
- Electric fields are always directed away from positive charges and toward negative charges.
- No matter how many charges are present, the number of lines per area passing perpendicular through a surface is proportional to the magnitude of the electric field.
- An **electric dipole** consists of two separated point charges that have the same magnitude but opposite signs.
- The electric field of a dipole is proportional to the product of the magnitude of one of the

charges and the distance between the charges. This product is called the **dipole moment**.
- Electric field lines always begin on a positive charge and end on a negative charge and do not start or stop in midspace. Furthermore, the number of lines leaving a positive charge or entering a negative charge is proportional to the magnitude of the charge.

18.8 THE ELECTRIC FIELD INSIDE A CONDUCTOR: SHIELDING

- At equilibrium under electrostatic conditions, any excess charge resides on the surface of a conductor.
- At equilibrium under electrostatic conditions, the electric field at any point within a conducting material is zero.
- The conductor **shields** any charge within it from electric fields created outside the conductor. The shielding results from the induced charges on the conductor surface.
- The electric field just outside the surface of a conductor is perpendicular to the surface at equilibrium under electrostatic conditions.

- The preceding material does not apply to insulators, which have very few free electrons.

18.9 GAUSS' LAW

- An extended collection of charges is called a **charge density**.
- A hypothetical closed surface is called a **Gaussian surface**.
- Gauss' law for a point charge is

$$EA = \frac{q}{\varepsilon_o} \qquad (18.5)$$

- In Gauss' law an important product is called the **electric flux**, Φ_E: $\Phi_E = EA$.
- A Gaussian surface can have any arbitrary shape, but must be closed.
- The direction of the electric field is not necessarily perpendicular to a Gaussian surface.
- If ϕ is the angle between the electric field and the normal

$$\Phi_E = \sum (E \cos\phi) \Delta A \qquad (18.6)$$

- **Gauss' Law**
  The electric flux Φ_E through a Gaussian surface is equal to the net charge Q enclosed by the surface divided by ε_o, the permittivity of free space:

  $$\sum (E\cos\phi)\Delta A = Q/\varepsilon_o \qquad (18.7)$$

- SI Unit of Electric Flux: $N\cdot m^2/C$
- The law is most useful when the distribution is uniform and symmetrical.
- The electric field of a charged thin spherical shell is $E = Q/(4\pi\varepsilon_o r^2)$ for $r > R$.
  The electric field inside a parallel plate capacitor is $E = \sigma/\varepsilon_o$.

18.10 COPIERS AND COMPUTER PRINTERS

- The copying process is called **xerography**, and is an application of static electricity. Other applications are the laser printer and the inkjet printer.

CHAPTER 19 | ***ELECTRIC POTENTIAL ENERGY AND THE ELECTRIC POTENTIAL***

This chapter introduces the study of electric potential, and the concept that there is an energy associated with electricity.

Capacitors are shown to store energy.

Important Concepts

- electric potential energy
- electric potential
- electron volt
- equipotential surface
- potential gradient
- capacitor
- dielectric constant
- energy density
- neuron
- dendrite
- axon
- synapse
- resting membrane potential
- action potential

19.1 POTENTIAL ENERGY

- The electrostatic force is conservative.
- The work done by the electric force equals the difference between the **electric potential energy** EPE at A and that at B:

$$W_{AB} = \text{EPE}_A - \text{EPE}_B \qquad (19.1)$$

19.2 THE ELECTRIC POTENTIAL DIFFERENCE

- Since the electric force $\mathbf{F} = q_o\mathbf{E}$, the work done by this force as the charge moves from A to B depends on the charge q_o. It is useful to express this work on a per-unit-charge basis as follows:

$$\frac{W_{AB}}{q_o} = \frac{\text{EPE}_A}{q_o} - \frac{\text{EPE}_B}{q_o} \qquad (19.2)$$

- **Definition of Electric Potential**

The **electric potential** V at a given point is the electric potential energy EPE of a small test charge q_o situated at that point divided by the charge itself:

$$V = \frac{\text{EPE}}{q_o} \qquad (19.3)$$

- SI Unit of Electric Potential: joule/coulomb = volt (V)
- The electric potential energy EPE and the electric potential V are not the same.
- The electric potential difference, $V_B - V_A$, between two points A and B is related to the work per charge in the following manner:

$$V_B - V_A = \frac{\text{EPE}_B}{q_o} - \frac{\text{EPE}_A}{q_o} = \frac{-W_{AB}}{q_o} \quad (19.4)$$

Often the "delta" notation is used to express the difference (final value minus initial value) in potentials and potential energies:

$$\Delta V = \frac{\Delta(\text{EPE})}{q_0} = \frac{-W_{AB}}{q_o} \qquad (19.4)$$

- Neither the potential V nor the potential energy EPE can be determined in an absolute sense, for only the differences ΔV and $\Delta(\text{EPE})$ are measurable in terms of the work W_{AB}.

- A positive charge accelerates from a region of higher electric potential energy (or higher potential) to a region of lower electric potential energy (or lower potential).
- A negative charge accelerates from a region of lower potential toward a region of higher potential.
- One **electron volt** is the change in potential energy of an electron ($q_o = 1.60 \times 10^{-19}$ C) when the electron moves through a potential difference of one volt. Thus,
$$1 \text{ eV} = 1.60 \times 10^{-19} \text{ J}$$

19.3 THE ELECTRIC POTENTIAL DIFFERENCE CREATED BY POINT CHARGES

- For a point charge $+q$

$$V_B - V_A = \frac{kq}{r_B} - \frac{kq}{r_A} \qquad (19.5)$$

- The potential of a point charge is

$$V = \frac{kq}{r} \qquad (19.6)$$

V refers to a potential difference with the arbitrary assumption that the potential at infinity is zero.
- When two or more charges are present, the potential due to all the charges is obtained by adding together the individual potentials.

19.4 EQUIPOTENTIAL SURFACES AND THEIR RELATION TO THE ELECTRIC FIELD

- An **equipotential surface** is a surface on which the electric potential is the same everywhere.
- The net electric force does no work as a charge moves on an equipotential surface.
- The electric field created by any group of charges is everywhere perpendicular to the associated equipotential surfaces and points in the direction of decreasing potential.
- The potential difference between capacitor plates can be written in terms of the electric field as $\Delta V = -W_{AB}/q_o = -q_o E \Delta s/q_o$, or

$$E = -\frac{\Delta V}{\Delta s} \qquad (19.7)$$

The quantity $\Delta V/\Delta s$ is referred to as the **potential gradient** and has the units of volts per meter.

19.5 CAPACITORS AND DIELECTRICS

- A **capacitor** consists of two conductors of any shape placed near one another without touching.
- An electrically insulating material used to fill the region between the conductors is called a **dielectric**.
- The proportionality $q \propto V$ can be expressed with the aid of a constant C that is called the **capacitance** of the capacitor.
- **The Relation Between Charge and Potential Difference for a Capacitor**
  The magnitude q of the charge on each plate of a capacitor is directly proportional to the magnitude V of the potential difference between the plates:

$$q = CV \qquad (19.8)$$

where C is capacitance.
- SI Unit of Capacitance: coulomb/volt = farad (F)
- Usually smaller amounts of capacitance, such as a microfarad or picofarad, are used in electric circuits.
- The capacitance reflects the ability of the capacitor to store charge, in the sense that a larger capacitance C allows more charge q to be put onto the plates for a given value of the potential difference V.
- If a dielectric is inserted between the plates of a capacitor, the capacitance can increase markedly.
- A dielectric utilizes the dipole nature of many materials.
- Because of the surface charges on the dielectric, not all the electric field lines generated by the charges on the plates pass through the dielectric. This reduction of the electric field is described by the **dielectric constant** κ, which is the ratio of the field magnitude E_o without the dielectric to the field magnitude E inside the dielectric:

$$\kappa = \frac{E_o}{E} \qquad (19.9)$$

- Being a ratio of two field strengths, the dielectric constant is a number without units.
- The value of κ depends on the nature of the material.
- For a parallel plate capacitor with a distance of d between plates of area A

$$C = \frac{\kappa \varepsilon_o A}{d} \qquad (19.10)$$

- Notice that only the geometry of the plates and the dielectric constant affect the capacitance.
- It can be shown that the relation $C = \kappa C_o$ applies to any capacitor, not just parallel plate capacitors.
- One reason that capacitors are filled with dielectric materials is to increase the capacitance.
- A capacitor is a device for storing charge.
- It is possible to view a capacitor as storing energy. The charge on the plates possesses electric potential energy, which arises because work was done to deposit the charges on the plates.
- The energy stored is

$$\text{Energy} = (1/2)(CV)V = (1/2)CV^2 \qquad (19.11)$$

- Since the area A times the separation d is the volume between the plates, the energy per unit volume or **energy density** is

$$\text{Energy density} = \frac{\text{Energy}}{\text{Volume}} = \tfrac{1}{2}\kappa\varepsilon_0 E^2 \quad (19.12)$$

It can be shown that this expression is valid for any electric field strength, not just that between the plates of a capacitor.

19.6 BIOMEDICAL APPLICATIONS OF ELECTRIC POTENTIAL DIFFERENCES

- A nerve consists of a bundle of nerve cells, and each cell is called a **neuron**.
- A neuron consists of a cell body to which numerous appendages known as **dendrites** are attached.
- Extending from the cell body is a long, thin trail called the **axon**, at the end of which are nerve endings.
- The axon transmits the signal to the nerve endings, which either stimulate a muscle cell or transmit the signal across a gap (called **synapse**) to the next neuron.

- Separation of positive and negative charges gives rise to an electric potential difference across the membrane, called the **resting membrane potential**.
- A significant change in the membrane potential away from and back to its normal resting value in a very short time is known at the **action potential**.

CHAPTER 20 | *ELECTRIC CIRCUITS*

This chapter introduces the study of electric circuits. Both direct and alternating current are considered. The concepts of electromotive force, current, resistance, and power are essential to understanding this material.

Ohm's law and the two Kirchhoff rules are of great importance.

Important Concepts

- battery
- electromotive force
- electric current
- ampere
- direct current
- alternating current
- conventional current
- resistance
- Ohm's law
- ohm
- resistor
- resistivity
- semiconductor
- temperature coefficient of resistivity
- critical temperature

- superconductor
- electric power
- kilowatt-hour
- root mean square current
- root mean square voltage
- series wiring
- parallel wiring
- short out
- internal resistance
- terminal voltage
- Kirchhoff's junction rule
- Kirchhoff's loop rule
- galvanometer
- ammeter
- shunt resistor
- voltmeter
- equivalent capacitance
- *RC* circuit
- time constant
- electrical grounding

20.1 ELECTROMOTIVE FORCE AND CURRENT

- Within a **battery**, a chemical reaction occurs that transfers electrons from one terminal (leaving it positively charged) to another terminal (leaving it negatively charged). Because of the positive and negative charges on the battery terminals, an electric potential difference exists between them. The maximum potential difference is called the **electromotive force** (*emf*) of the battery, for which the symbol ξ is used.

- In reality, the potential difference between the terminals of a battery is somewhat less than the maximum value indicated by the emf.
- The battery creates an electric field within and parallel to the wire, directed from the positive toward the negative terminal The field exerts a force on the free electrons in the wire, and they respond by moving. The resulting flow of charge is known as an **electric current**.
- The current is defined as the amount of charge per unit time that crosses an imaginary surface that is perpendicular to their motion.
- If the rate of flow of charge is constant, the current I is

$$I = \frac{\Delta q}{\Delta t} \qquad (20.1)$$

If the rate is not constant, then Equation 20.1 gives the average current.

- Since the units for charge and time are the coulomb (C) and the second (s), the SI unit for current is the coulomb per second (C/s).
- One coulomb per second is referred to as an **ampere** (A).
- If the charges move around a circuit in the same direction at all times, the current is said to be **direct current** (dc).
- The current is said to be **alternating current** (ac) when the charges move first one way and then the opposite way.
- Electrons flow in metal wires.
- It is customary not to use the flow of electrons when discussing circuits. Instead, a so-called **conventional current** is used. Conventional current is the hypothetical flow of positive charge that would have the same effect in a circuit as the movement of negative charges that actually does occur.
- In this text the symbol I stands for conventional current.

20.2 OHM'S LAW

- The **resistance** (R) is defined as the ratio of the voltage V applied across a piece of material, or $R = V/I$.
- **Ohm's Law**
  The ratio V/I is a constant, where V is the voltage applied across a piece of material (such as a wire) and I is the current through the material:

$$\frac{V}{I} = R = \text{constant or } V = IR \qquad (20.2)$$

  R is the resistance of the piece of material.
  SI Unit of Resistance: volt/ampere (V/A) = ohm (Ω).
- Ohm's law is not a fundamental law of nature.
- To the extent that a wire or an electrical device offers resistance to the flow of charges, it is called a **resistor**.
- A zigzag line represents a resistor and a straight line represents an ideal conducting wire, or one with negligible resistance.

20.3 RESISTANCE AND RESISTIVITY

- For a wide range of materials, the resistance of a piece of material of length L and cross-sectional area A is

$$R = \rho \frac{L}{A} \qquad (20.3)$$

where ρ is a proportionality constant known as the **resistivity** of the material.
- The unit for resistivity is the ohm·meter ($\Omega \cdot$m).
- Conductors have small resistivities.
- Insulators have large resistivities.
- Materials with intermediate resistivities are called **semiconductors**.
- Resistance depends on both the resistivity and the geometry of the material.
- For many materials and limited temperature ranges, it is possible to express the temperature dependence of the resistivity as follows:

$$\rho = \rho_o [1 + \alpha (T - T_o)] \qquad (20.4)$$

In this expression ρ and ρ_o are the resistivities at temperatures T and T_o, respectively.

- The term α has the unit of reciprocal temperature and is the **temperature coefficient of resistivity**.
- When resistivity increases with increasing temperature, α is positive.
- When resistivity decreases with increasing temperature, α is negative.
- Resistance depends on temperature according to

$$R = R_o[1 + \alpha(T - T_o)] \qquad (20.5)$$

- There is an important class of materials whose resistivity suddenly goes to zero below a certain temperature T_c called the **critical temperature**, commonly a few degrees above absolute zero. Below this temperature, such materials are called **superconductors**.
- Recently, ceramic materials have been developed that undergo the transition to the superconducting state at much higher temperatures.

20.4 ELECTRIC POWER

- **Electric Power**
  When there is a current I in a circuit as a result of a voltage V, the electric power P delivered to the circuit is

- $$P = IV \qquad (20.6)$$

- SI Unit of Power: watt (W)

- $$P = IV \qquad (20.6a)$$

- $$P = I(IR) = I^2 R \qquad (20.6b)$$

- $$P = \left(\frac{V}{R}\right)V = \frac{V^2}{R} \qquad (20.6c)$$

- A commonly used unit for energy is the **kilowatt-hour** (kWh).

20.5 ALTERNATING CURRENT

- The most common type of ac voltage fluctuates sinusoidally as a function of time t:

$$V = V_o \sin 2\pi ft \qquad (20.7)$$

- In a circuit containing only resistance, the ac current also fluctuates sinusoidally as a function of time:

$$I = (V_o/R) \sin 2\pi ft = I_o \sin 2\pi ft \qquad (20.8)$$

- The peak current is given by $I_o = V_o/R$, so it can be determines if the peak voltage and resistance are known.
- The power delivered to an ac circuit is

$$P = I_o V_o \sin^2 2\pi ft \qquad (20.9)$$

- Since the power fluctuates in an ac circuit, it is customary to consider the average power $\overline{P}$, which is one-half the peak power:

$$\overline{P} = \frac{1}{2} I_o V_o \qquad (20.10)$$

$$\overline{P} = \left(\frac{I_o}{\sqrt{2}}\right)\left(\frac{V_o}{\sqrt{2}}\right) = I_{rms} V_{rms} \qquad (20.11)$$

- I_{rms} and V_{rms} are called the **root mean square (rms) voltage and current**.

$$I_{rms} = \frac{I_o}{\sqrt{2}} \qquad (20.12)$$

$$V_{rms} = \frac{V_o}{\sqrt{2}} \qquad (20.13)$$

- When an ac voltage or current is specified in this text, it is an rms value, unless indicated otherwise.
- Ohm's law can be written conveniently in terms of rms quantities:

$$V_{rms} = I_{rms} R \qquad (20.14)$$

- Average power can be expressed in the following way:

$$\overline{P} = I_{rms} V_{rms} \qquad (20.15a)$$

$$\overline{P} = I_{rms}^2 R \qquad (20.15b)$$

$$\overline{P} = \frac{V_{rms}^2}{R} \qquad (20.15c)$$

20.6 SERIES WIRING

- **Series wiring** means that the devices are connected in such a way that there is the same electric current through each device.
- For **series resistors**, the equivalent resistance is

$$R_S = R_1 + R_2 + R_3 + \ldots \qquad (20.16)$$

20.7 PARALLEL WIRING

- **Parallel wiring** means that the electric devices are connected in such a way that the same voltage is applied across each device.
- Two parallel resistors have a single equivalent resistance that is smaller than either resistor.

- For **parallel resistors**, the equivalent resistance R_P can be obtained from the following equation:

$$\frac{1}{R_P} = \frac{1}{R_1} + \frac{1}{R_2} + \frac{1}{R_3} + \ldots \quad (20.17)$$

- In a parallel combination of resistances, the smallest resistance has the largest impact in determining the equivalent resistance.
- A near-zero resistance is said to **short out** the other resistances, by providing a near-zero resistance path for the current to follow as a shortcut around the other parallel resistances.

20.8 CIRCUITS WIRED PARTIALLY IN SERIES AND PARTIALLY IN PARALLEL

- Often an electric circuit is wired partially in series and partially in parallel. The key to determining the current, voltage, and power in such a case is to deal with the circuit in parts, with resistances in each part being either in series or in parallel with each other.

20.9 INTERNAL RESISTANCE

- The resistance added by a battery or a generator is called the **internal resistance**.
- When current is drawn from a battery, the internal resistance causes the voltage between the terminals to drop below the maximum value specified by the battery's emf. The actual voltage between the terminals of a battery is known as the **terminal voltage**.

20.10 KIRCHHOFF'S RULES

- **Kirchhoff's Rules**
  Junction rule. The sum of the magnitudes of the currents directed into a junction equals the sum of the magnitudes of the currents directed out of the junction.
  Loop rule. Around any closed circuit loop, the sum of the potential drops equals the sum of the potential rises.

20.11 THE MEASUREMENT OF CURRENT AND VOLTAGE

- A **galvanometer** consists of a magnet, a coil of wire, a spring, a pointer, and a calibrated scale. The coil rotates in response to the torque applied by the magnet when there is a current in the coil.
- The amount of dc current that causes full-scale deflection of the pointer indicates the sensitivity of the galvanometer.
- An **ammeter** is an instrument that measures current. It must be inserted in the circuit so the current passes directly through it. An ammeter includes a galvanometer and one or more **shunt resistors**, which are connected in parallel with the galvanometer and provide a bypass for current in excess of the galvanometer's full-scale limit.
- An ideal ammeter would have zero resistance.
- A **voltmeter** is an instrument that measures the voltage between two points.
- A voltmeter must be connected between the points and is not inserted into the circuit as an ammeter is.

- A voltmeter includes a galvanometer whose scale is calibrated in volts.
- Ideally, the voltage measured by a voltmeter should be the same voltage that exists when the voltmeter is not connected. However, a voltmeter takes some current from a circuit and, thus, alters the circuit voltage to some extent.
- An ideal voltmeter would have an infinite resistance and draw away only an infinitesimal amount of current.

20.12 CAPACITORS IS SERIES AND PARALLEL

- For parallel capacitors, the equivalent capacitance is

$$C_P = C_1 + C_2 + C_3 + \ldots \qquad (20.18)$$

- The total energy stored in the equivalent capacitor C_P is

$$\text{Total energy} = (1/2)C_P V^2$$

- All capacitors in series, regardless of their capacitances, contain charges of the same magnitudes, $+q$ and $-q$, on their plates.
- The equivalent capacitance of series capacitors can be obtained from the following equation:

$$\frac{1}{C_s} = \frac{1}{C_1} + \frac{1}{C_2} + \frac{1}{C_3} + \ldots \qquad (20.19)$$

20.13 RC CIRCUITS

- A circuit that contains both resistors and capacitors is called an **RC circuit**.
- It can be shown that the magnitude q of charge on the plates of a capacitor at time t is

$$q = q_o[1 - e^{-t/(RC)}] \qquad (20.20)$$

assuming that the capacitor is uncharged at time $t = 0$.

- The term RC in the exponent in Equation 20.20 is called the **time constant** τ of the circuit:

$$\tau = RC \qquad (20.21)$$

- The time constant is measured in seconds.
- The time constant is to the time required for an uncharged capacitor to accumulate 63.2% of its equilibrium charge.
- Assuming the capacitor has a charge q_o at a time $t = 0$ when the switch is closed, it can be shown that

$$q = q_o e^{-t/(RC)} \qquad (20.22)$$

where q is the amount of charge remaining on either plate at time t.

20.14 SAFETY AND THE PHYSIOLOGICAL EFFECTS OF CURRENT

- Proper **grounding** is necessary.
- Currents on the order of 0.01-0.02 A can lead to muscle spasms, in which the person "can't let go" of the object causing the shock. Currents of approximately 0.2A are potentially fatal.

CHAPTER 21 | *MAGNETIC FORCES AND MAGNETIC FIELDS*

This chapter shows the relationship between magnetism and electricity by discussing the force that a magnetic field applies to moving electric charges or electric currents and the magnetic fields created by electric currents.

Electric motors are also discussed.

Important Concepts

- magnetic pole
- magnetic field
- magnetic field lines
- angle of declination
- angle of dip
- magnetic force
- right-hand-rule no. 1
- tesla
- gauss
- bubble chamber
- mass spectrometer
- magnetic moment
- direct-current motor
- armature

- half-ring
- right-hand rule no. 2
- permeability of free space
- solenoid
- ampere's law
- ferromagnetic material
- magnetic domain
- fringe field

21.1 MAGNETIC FIELDS

- Permanent magnets have long been used in navigational compasses. The end of the compass needle that points north is labeled the **north magnetic pole**; the opposite end is the **south magnetic pole**.
- Like poles repel each other, and unlike poles attract.
- No one has found a magnetic monopole (an isolated north or south pole).
- Surrounding a magnet, there is a **magnetic field**.
- The magnetic field has both magnitude and direction.
- The direction of the magnetic field at any point in space is the direction indicated by the north pole of a small compass needle placed at that point.

- It is possible to draw **magnetic field lines** in the vicinity of a magnet. The lines appear to originate from the north pole and to end on the south pole, the lines do not start or stop in midspace.
- The magnetic field at any point is tangent to the magnetic field line at that point. Furthermore, the strength of the magnetic field is proportional to the number of lines per unit area that passes through a surface oriented perpendicular to the lines.
- The lines are closest together near the north and south poles, reflecting the fact that the strength of the magnetic field is greatest in these regions.
- The location where the magnetic axis crosses the earth's surface in the northern hemisphere is known as the north magnetic pole. The north magnetic pole is so named because it is the location toward which the north end of a compass needle points.
- The north magnetic pole does not coincide with the north geographic pole. The position of the north magnetic pole is not fixed, but moves over the years.
- The angle that a compass needle deviates from the north geographic pole is called the **angle of declination**.

- The angle that the earth's magnetic field makes with respect to the earth's surface at any point is known as the **angle of dip**.

21.2 THE FORCE THAT A MAGNETIC FIELD EXERTS ON A MOVING CHARGE

- A charge placed in a magnetic field experiences a **magnetic force** provided two conditions are met:
  1. The charge must be moving, for no magnetic force acts on a stationary charge.
  2. The velocity of the moving charge must have a component that is perpendicular to the direction of the magnetic field.
- If a charge moves parallel or antiparallel to the field, the charge experiences no magnetic force.
- If a charge moves perpendicular to the field, the charge experiences a maximum force.
- **Right-Hand Rule No. 1.** Extend the right hand so the fingers point along the direction of the magnetic field **B** and the thumb points along the velocity **v** of the charge. The palm of the hand then faces in the direction of the magnetic force **F** that acts on a positive charge.

- If the moving charge is negative instead of positive, the direction of the magnetic force is opposite to that predicted by RHR-1.
- **Definition of the Magnetic Field**
  The magnitude B of the magnetic field at any point is defined as

  $$B = \frac{F}{q_o(v\sin\theta)} \quad (21.1)$$

  where F is the magnitude of the magnetic force on a positive test charge q_o and v is the velocity of the charge and makes an angle θ ($0 \le \theta \le 180°$) with the direction of the magnetic field. The magnetic field **B** is a vector, and its direction can be determined by using a small compass needle.
- SI Unit of Magnetic Field: $\frac{\text{newton} \cdot \text{second}}{\text{coulomb} \cdot \text{meter}} =$ 1 **tesla** (T)
- One tesla is the strength of the magnetic field in which a unit test charge, traveling perpendicular to the magnetic field with a speed of one meter per second, experiences a force of one newton. Because a coulomb per second is an ampere, the tesla is often written as 1 T = 1 N/(A·m).
- 1 **gauss** = 10^{-4} tesla

21.3 THE MOTION OF A CHARGED PARTICLE IN A MAGNETIC FIELD

- There is no work done by a static magnetic field on a moving charged particle. This fact arises because the magnetic force always acts in a direction that is perpendicular to the motion of the charge.
- The magnetic force always remains perpendicular to the velocity and is directed toward the center of the circular path.

$$r = \frac{mv}{qB} \qquad (21.2)$$

- A **bubble chamber** contains a superheated liquid such as hydrogen, which will boil and form bubbles readily. When an electrically charged particle passes through the chamber, a thin track of bubbles is left in its wake.

21.4 THE MASS SPECTROMETER

- In one type of **mass spectrometer**, the atoms of molecules are first vaporized and then ionized by the ion source. The ions pass through a hole in a plate and enter a region of constant magnetic field **B**, where they are deflected in semicircular paths. Only those ions with the proper radius r strike the detector.

21.5 THE FORCE ON A CURRENT IN A MAGNETIC FIELD

- A current in the presence of a magnetic field can also experience a force.
- The direction of the force is determined by using the RHR-1, with the minor modification that the direction of the velocity of a positive charge is replaced by the direction of the conventional current I.
- The magnetic force on a current-carrying wire of length L is

$$F = ILB \sin \theta \qquad (21.3)$$

- The force on a current-carrying wire is maximum when the wire is oriented perpendicular to the field and vanishes when the current is parallel or antiparallel to the field.

21.6 THE TORQUE ON A CURRENT-CARRYING COIL

- If a loop of wire is suspended properly in a magnetic field, the magnetic force produces a torque that tends to rotate the loop. This torque is responsible for the operation of a widely used type of electric motor.
- The torque is maximum when the normal to the plane of the loop is perpendicular to the field. The torque is zero when the normal is parallel to the field. When a current-carrying loop is placed in a magnetic field, the loop tends to rotate such that its normal becomes aligned with the magnetic field.
- If a wire is wrapped so as to form a coil containing N loops the torque becomes

$$\tau = NIAB \sin \phi \qquad (21.4)$$

- The torque depends on
  (1) the geometric properties of the coil itself and the current in it.
  (2) the magnitude B of the magnetic field.
  (3) the orientation of the normal to the coil with respect to the direction of the field.

- The quantity NIA is known as the **magnetic moment** of the coil, and its units are ampere·meter^2.

- A **direct current** (dc) **motor** consists of a coil of wire placed in a magnetic field and free to rotate about a vertical shaft. The coil of wire contains many turns and is wrapped around a cylinder that rotates with the coil. The coil and iron cylinder assembly is known as the **armature**. Each end of the wire coil is attached to a metallic **half-ring**.

- A current-carrying wire produced a magnetic field. The magnetic field produced by the current are circles centered on the wire.

21.7 MAGNETIC FIELDS PRODUCED BY CURRENTS

- **Right-Hand Rule No. 2**
  Curl the fingers of the right hand into the shape of a half-circle. Point the thumb in the direction of the conventional current I, and the tips of the fingers will point in the direction of the magnetic field **B**.
- The magnetic field produced by a current I in an infinitely long wire at a distance r from the wire is

$$B = \frac{\mu_o I}{2\pi r} \qquad (21.5)$$

- The constant μ_o is known as the **permeability of free space**, and its value is $\mu_o = 4\pi \times 10^{-7}$ T·m/A.
- If the current-carrying wire is bent into a circular loop, the magnetic field at the center of the loop of radius R is perpendicular to the plane of the loop and has the value $B = \mu_o I/(2R)$, where I is the current in the loop. Often the loop consists of N turns of wire that are wound sufficiently close together and produce a magnetic field

$$B = N\frac{\mu_o I}{2R} \qquad (21.6)$$

- The direction of the magnetic field at the center of a loop can be determined with the help of RHR-2.
- The loop itself behaves as a bar magnet with a "north pole" on one side and a "south pole" on the other side.
- A **solenoid** is a long coil of wire in the shape of a helix.
- The magnitude of the magnetic field in the interior of a long solenoid is

$$B = \mu_o nI \qquad (21.7)$$

where n is the number of turns per unit length of the solenoid and I is the current.

21.8 AMPERE'S LAW

- **Ampere's Law for Static Magnetic Fields**
  For any current geometry that produces a magnetic field that does not change in time,

$$\sum B_{\parallel} \Delta \ell = \mu_o I \qquad (21.8)$$

where $\Delta \ell$ is a small segment of length along a closed path of arbitrary shape around the current. $B_{\parallel}$ is the component of the magnetic field parallel to $\Delta \ell$. I is the net current passing through a surface bounded by the path, and μ_o is the permeability of free space. The symbol $\sum$ indicates that the sum of all $B_{\parallel}\Delta \ell$ terms must be taken around the closed path.

21.9 MAGNETIC MATERIALS

- The magnetic field around a bar magnet is also due to the motion of charges. The motion responsible for the magnetism is that of the electrons within the atoms of the material.
- First, each electron orbiting the nucleus behaves like an atomic-sized loop of current that generates a small magnetic field.
- Second, each electron possesses a spin that also gives rise to a magnetic field.
- In most substances the magnetism produced at the atomic level tends to cancel out, with the

result that the substance is nonmagnetic overall.
- There are some materials, known as **ferromagnetic materials**, in which the cancellation does not occur for groups of approximately 10^{16}-10^{19} neighboring atoms, because they have electron spins that are naturally aligned parallel to each other.
- Such a region is called a **magnetic domain**.
- Often the magnetic domains in a ferromagnetic material are arranged randomly.
- The magnetic domains can be aligned with the aid of an external magnetic field. They remain aligned for the most part after the magnetic field is removed, and the material becomes permanently magnetized.
- It is not unusual for the induced magnetic field to be a hundred to a thousand times stronger than the external field that causes the alignment.
- A permanent magnet attracts a piece of iron by inducing an opposite magnetic pole in the side of the iron nearest the magnet.

21.10 OPERATIONAL DEFINITIONS OF THE AMPERE AND THE COULOMB

- Suppose that the same current I is sent through two long, straight, parallel wires that are separated by a distance r. The magnetic field produced by the current in one wire is perpendicular to the wire and exerts a force on the other wire of:

- $$F = ILB\sin 90° = \frac{\mu_o I^2 L}{2\pi r} \qquad (21.3)$$

  This expression can be used to define what is meant by a current of one ampere.
- One coulomb is defined as that quantity of electric charge that passes a given point in one second when the current is one ampere.

- $\mu_o = 4\pi \times 10^{-7}$ T·m/A.

CHAPTER 22 | ***ELECTROMAGNETIC INDUCTION***

This chapter discusses induction in the presence of a magnetic field. The induction can be a result of the field changing or the conductor moving.

It is necessary to consider magnetic flux and the orientation of the magnetic field with respect to the motion to discuss induction fully.

The transformer is a significant application of electromagnetic induction.

Important Concepts

- induced current
- induced emf
- electromagnetic induction
- motional emf
- magnetic flux
- weber
- Faraday's law of electromagnetic induction
- induced magnetic field
- Lenz's law
- electric generator
- alternating current generator
- load

- countertorque
- back emf
- counter emf
- primary coil
- secondary coil
- mutual induction
- henry
- self-induction
- self-inductance
- energy density
- transformer
- transformer equation
- step-up
- step-down
- turns ratio

22.1 INDUCED EMF AND INDUCED CURRENT

- The current in a coil is called an **induced current**, if it is brought about by a changing magnetic field.
- Since a source of emf is always needed to produce a current, the coil itself behaves as if it

were a source of emf. This emf is known as an **induced emf**.

- The phenomenon of producing an induced emf with the aid of a magnetic field is called **electromagnetic induction**.

22.2 MOTIONAL EMF

- When a conducting rod moves through a constant magnetic field, an emf is induced in the rod. This special case of electromagnetic induction arises as a result of the magnetic force that acts on a moving charge.
- The separated charges on the ends of the moving conductor give rise to an induced emf, called a **motional emf**.
- The motional emf when v, B, and L are mutually perpendicular is

$$\xi = vBL \qquad (22.1)$$

- As expected, $\xi = 0$ when $v = 0$, for no motional emf is developed in a stationary rod.
- Motional emf arises because a magnetic force acts on the charges in a conductor that is moving through a magnetic field. Whenever

this emf causes a current, a second magnetic force enters the picture. The second force **F** arises because the current I in the rod is perpendicular to the magnetic field.
- The direction of **F** is specified by RHR-1 and is opposite the velocity **v** of the rod. By itself, **F** would slow down the rod.

22.3 MAGNETIC FLUX

- **Magnetic flux** is analogous to electric flux.
- The general expression for magnetic flux through a surface of area A is

$$\Phi = (B \cos \phi)A = BA \cos \phi \qquad (22.2)$$

where B is the magnitude of the magnetic field and ϕ is the angle between the field and the normal to the surface.
- If either the magnitude B of the magnetic field or the angle ϕ is not constant over the surface, an average value of the product $B \cos \phi$ must be used to compute the flux.
- The unit of magnetic flux is the tesla·meter^2 (T·m^2). This unit is called a **weber** (Wb).

- The magnetic flux is proportional to the number of field lines that passes through a surface.

22.4 FARADAY'S LAW OF ELECTROMAGNETIC INDUCTION

- Faraday discovered that whenever there is a change in flux through a loop of wire, an emf is induced in the loop. In this context, the word "change" refers to a change as time passes. A flux that is constant in time creates no emf.
- **Faraday's Law of Electromagnetic Induction** The average emf ξ induced in a coil of N loops is

$$\xi = -N\left(\frac{\Phi - \Phi_o}{t - t_o}\right) = -N\frac{\Delta\Phi}{\Delta t} \qquad (22.3)$$

where $\Delta\Phi$ is the change in magnetic flux through one loop and Δt is the time interval during which the change occurs. The term $\Delta\Phi/\Delta t$ is the average time rate of change of the flux that passes through one loop.
- SI Unit of Induced Emf: volt (V).

22.5 LENZ'S LAW

- The field created by the induced current is called the **induced magnetic field**.
- **Lenz's Law**
  The induced emf resulting from a changing magnetic flux has a polarity that leads to an induced current whose direction is such that the induced magnetic field opposes the original flux change.
- Lenz's law should not be thought of as an independent law, for it is a consequence of the conservation of energy.

22.6 APPLICATIONS OF ELECTROMAGNETIC INDUCTION TO THE REPRODUCTION OF SOUND

- Electromagnetic induction plays an important role in the technology used for the reproduction of sound.

- The playback head of a cassette deck uses a moving tape to generate an emf in a coil of wire.
- There are microphones known as moving coil microphones.

22.7 THE ELECTRIC GENERATOR

- **Electric generators** produce virtually all of the world's electrical energy. A generator produces electrical energy from mechanical work, just the opposite of what a motor does. In a motor, an input electric current causes a coil to rotate, thereby doing mechanical work on any object attached to the shaft of the motor. In a generator, the shaft is rotated by some mechanical means, such as an engine or turbine, and an emf is induced in a coil.
- In its simplest form, an ac generator consists of a coil of wire that is rotated in a uniform magnetic field.
- The emf induced in a rotating planar coil containing N turns is

$$\xi = NAB\omega \sin \omega t = \xi_o \sin \omega t \text{ where } \omega = 2\pi f$$
(22.4)

In this result the angular speed ω is in radians per second and is related to the frequency f.
- The result is valid for any planar shape of area A and shows that the emf varies sinusoidally with time.
- This electric generator is also called an **alternating current (ac) generator**.
- The devices to which the generator supplies electricity are known collectively as the "**load**."
- A magnetic force gives rise to a **countertorque** that opposes the rotational motion. The greater the current drawn from a generator, the greater the countertorque, and the harder it is for the turbine to turn the coil.
- A generator converts mechanical energy to electrical energy: in contrast, an electric motor converts electrical energy to mechanical energy.
- In an electric motor, the induced emf that acts to oppose the applied emf is called a **back emf** or **counter emf**.
- If R is the resistance of the wire in the coil, and the current I drawn by the motor is determined from Ohm's law as the net emf divided by the resistance, we have

$$I = \frac{V - \xi}{R} \qquad (22.5)$$

22.8 MUTUAL INDUCTANCE AND SELF-INDUCTANCE

- The **primary coil** is the one connected to an ac generator, which sends an alternating current I_1 through it.
- The **secondary coil** is not attached to a generator, although a voltmeter connected across it will register an induced emf.
- The effect in which a changing current in one circuit induces an emf in another circuit is called **mutual induction**.
- In the following equation, M is known as the **mutual inductance.**

$$N_2\Phi_2 = MI_1 \text{ or } M = \frac{N_2\Phi_2}{I_1} \quad (22.6)$$

- The emf due to mutual induction is

$$\xi_2 = -M\frac{\Delta I_1}{\Delta t} \quad (22.7)$$

- The measurement unit for the mutual inductance M is V·s/A, which is called a **henry**.

- The mutual inductance depends on the geometry of the coils and the nature of any ferromagnetic material present.
- An alternating current creates an alternating magnetic field that, in turn, creates a changing flux through a coil. The change in flux induces an emf in the coil, in accord with Faraday's law. The effect in which a changing current in a circuit induces an emf in the same circuit is referred to as **self-inductance**.

$$N\Phi = LI \text{ or } L = \frac{N\Phi}{I} \qquad (22.8)$$

The emf due to self-induction is

$$\xi = -L\frac{\Delta I}{\Delta t} \qquad (22.9)$$

- Like mutual inductance, L is measured in henries.
- An inductor, like a capacitor, can store energy. The stored energy arises because a generator does work to establish a current in an inductor.
- The stored energy is

$$\text{Energy} = \frac{1}{2}LI^2 \qquad (22.10)$$

- The term $A\ell$ is the volume inside a solenoid where the magnetic field exists, so the energy per unit volume or **energy density** is

  Energy density = Energy/Volume =
  $$\frac{1}{2\mu_o}B^2 \qquad (22.11)$$

- Although Equation 22.11 was obtained for the special case of a long solenoid, it is quite general and is valid for any point where a magnetic field exists in air or a vacuum or in a nonmagnetic material. Thus, energy can be stored in a magnetic field.

22.9 TRANSFORMERS

- A **transformer** is a device for increasing or decreasing an ac voltage.
- A typical transformer consists of an iron core on which two coils are wound: a **primary coil** with N_P turns, and a **secondary coil** with N_S turns. The primary coil is connected to the ac generator.
- The alternating current in the primary coil establishes a changing magnetic field in the iron core. Because iron is easily magnetized, it

greatly enhances the magnetic field relative to that in an air core and guides the field lines to the secondary coil. Since the magnetic field is changing, the flux through the primary and secondary coils is also changing, and consequently an emf is induced in both coils.
- The **transformer equation** is usually written in terms of terminal voltages V_S and V_P for the secondary and primary coils:

$$\frac{V_S}{V_P} = \frac{N_S}{N_P} \qquad (22.12)$$

- If N_S is greater that N_P the transformer is a **step-up** transformer.
- If N_S is less that N_P the transformer is a **step-down** transformer.
- The ratio N_S/N_P is referred to as the **turns ratio**.

If the power transferred is 100%

$$I_S/I_P = V_P/V_S = N_P/N_S \qquad (22.13)$$

- A transformer that steps up the voltage simultaneously steps down the current, and a transformer that steps down the voltage steps up the current.

CHAPTER 23 | ***ALTERNATING CURRENT CIRCUITS***

The material in this chapter is concentrated on alternating current in circuits. In particular, this chapter considers the total impedance, and how phase angles are handled when capacitors and inductors are present.

Resonance is a main concern of this chapter.

The chapter also discusses semiconductors, in particular, diodes and transistors.

Important Concepts

- capacitive reactance
- inductive reactance
- series RCL circuit
- impedance
- power factor
- resonance
- doping
- n-type semiconductor
- p-type semiconductor
- p-n junction
- forward bias
- reverse bias

- rectifier circuit
- bipolar junction transistor
- pnp transistor
- npn transistor

23.1 CAPACITORS AND CAPACITIVE REACTANCE

- The rms voltage across a capacitor is:

- $$V_{rms} = I_{rms} X_C \tag{23.1}$$

  The term X_C appears in place of the resistance R and is called the **capacitive reactance**.
- The capacitive reactance is measured in ohms.
- It is found experimentally that the capacitive reactance X_C is inversely proportional to both the frequency f and the capacitance C, according to the following equation:

$$X_C = \frac{1}{2\pi fC} \tag{23.2}$$

- The current in a resistor is *in phase* with the voltage across the resistor.
- The current in a capacitor leads the voltage across the capacitor by a phase angle of 90°.

- In an ac circuit, a capacitor uses no average power.
- The voltage and the current are one-quarter wave cycle out of phase.
- Since the current reaches its maximum value before the voltage across the capacitor does, it is said that the current in a capacitor *leads* the voltage across the capacitor by a phase angle of 90°.
- The capacitor alternately absorbs and releases energy. On the average, the power is zero and a capacitor uses no energy in an ac circuit.

23.2 INDUCTORS AND INDUCTIVE REACTANCE

- The rms voltage across an inductor is

$$V_{rms} = I_{rms}X_L \qquad (23.3)$$

The term X_L is called the **inductive reactance**.
- The inductive reactance is measured in ohms. It is found experimentally to be directly proportional to the frequency f and to the inductance L, according to

$$X_L = 2\pi f L \qquad (23.4)$$

- For an inductor, the current and the voltage are not in phase but are one-quarter of a wave cycle out of phase.
- The current reaches its maximum after the voltage does, and it is said that the current *lags behind* the voltage by a phase angle of 90°.
- An inductor alternately absorbs and releases energy. On the average, the power is zero and the inductor uses no energy in an ac circuit.

23.3 CIRCUITS CONTAINING RESISTANCE, CAPACITANCE, AND INDUCTANCE

- In a series **RCL** circuit the total opposition to the flow of charges is called the **impedance** of the circuit and comes partially from (1) the resistance R, (2) the capacitive reactance X_C, and (3) the inductive reactance X_L.
- The applied rms voltage is related to the voltages across each device according to

$$V_{rms}^2 = V_R^2 + (V_L - V_C)^2 \qquad (23.5)$$

- For the entire RCL circuit, it follows that

$$V_{rms} = I_{rms}Z \qquad (23.6)$$

where the impedance Z of the circuit is defined as

$$Z = \sqrt{R^2 + (X_L - X_C)^2} \qquad (23.7)$$

- The impedance of a circuit is measured in ohms.
- The phase angle for a series RCL combination is the angle ϕ between the current phasor I_o and the voltage phasor V_o.

$$\tan\phi = \frac{X_L - X_C}{R} \qquad (23.8)$$

- Only the resistor consumes power.

$$\overline{P} = I_{rms}V_{rms}\cos\phi \qquad (23.9)$$

where V_{rms} is the rms voltage of the generator. The term $\cos\phi$ is called the **power factor** of the circuit.

23.4 RESONANCE IN ELECTRIC CIRCUITS

- The behavior of the current and voltage in a series RCL circuit can give rise to a condition of **resonance**. Resonance occurs when the frequency of a vibrating force exactly matches a natural (resonant) frequency of the object to which the force is applied.
- A condition of resonance can be established in a series RCL circuit. In this case there is only one natural frequency, and the vibrating force is provided by the oscillating electric field that is related to the voltage of the generator.
- An ac circuit can have a resonant frequency f_0, because there is a natural tendency for energy to shuttle back and forth between the electric field of the capacitor and the magnetic field of the inductor.

$$f_0 = 1/(2\pi\sqrt{LC}) \qquad (23.10)$$

- The resonant frequency is determined by the inductance and the capacitance, but not the resistance.

- The effect of resistance on electrical resonance is to make the "sharpness" of the circuit response less pronounced.

23.5 SEMICONDUCTOR DEVICES

- Semiconductors are not pure materials, because small amounts of "impurity" atoms have been added to them to change their conductive properties.
- The process of adding impurity atoms is called **doping**.
- A semiconductor doped with an impurity that contributes mobile electrons is called an **n-type semiconductor**.
- A semiconductor doped with an impurity that introduces mobile positive holes is called a **p-type semiconductor**.
- A **p-n junction diode** is a device formed from a p-type semiconductor and an n-type semiconductor. The p-n junction between the two materials is of fundamental importance to the operation of diodes and transistors.
- Suppose a battery is connected across a p-n junction with the positive terminal attached to

- the p-material. In this situation the junction is said to be in a condition of **forward bias**.
- If the battery polarity is reversed, the p-n junctions in a condition known as **reverse bias**.
- The direction of the arrowhead in the diode symbol indicates the direction of the conventional current in the diode under a forward bias condition.
- Because diodes are unidirectional devices, they are commonly used in **rectifier circuits**, which convert an ac voltage into a dc voltage.
- Solar cells use p-n junctions to convert sunlight into electricity. The outer covering of p-type material is so thin that sunlight penetrates into the charge layers and ionizes some of the atoms there. In the process of ionization, the energy of the sunlight causes a negative electron to be ejected from the atom, leaving behind a positive hole.
- A **bipolar junction transistor** consists of two p-n junctions formed by three layers of doped semiconductors.
- A transistor is useful because it can be used in circuits that amplify a smaller voltage into a larger one.
- Transistors may be **pnp or npn type transistors**.

CHAPTER 24 | *ELECTROMAGNETIC WAVES*

The chapter includes a study of electromagnetic waves. Electromagnetic waves carry energy, even through a vacuum, and all electromagnetic waves travel at the speed of light in a vacuum.

The chapter discusses the amount of energy carried. If also goes into the concepts of the Doppler effect and polarization.

Important Concepts

- near field
- radiation field
- speed of light in a vacuum
- electromagnetic spectrum
- energy density
- intensity
- Doppler effect
- linearly polarized
- Polaroid
- transmission axis
- polarizer
- analyzer

- Malus' law
- liquid crystal display (LCD)

24.1 THE NATURE OF ELECTROMAGNETIC WAVES

- Both electric and magnetic fields are created by an antenna. Those that exist mainly near the antenna together are called the **near field**.
- Electric and magnetic fields do form a wave at large distances from an antenna. These fields together are referred to as the **radiation field**.
- An electromagnetic wave is a transverse wave, because the electric and magnetic fields are both perpendicular to the direction in which the wave travels.
- This kind of transverse wave does not require a medium in which to propagate. Electromagnetic waves can travel through a vacuum or a material substance, since electric and magnetic fields can exist in either one.
- In general, any electric charge that is accelerating emits an electromagnetic wave, whether the charge is inside a wire or not.
- All electromagnetic waves move through a vacuum at the same speed, and the symbol c is used to denote its value. This speed is called

the **speed of light in a vacuum** and is $c = 3.00 \times 10^8$ m/s.
- In air, electromagnetic waves travel at nearly the same speed as they do in a vacuum, but, in general, they move through a substance such as glass at a speed that is less than c.
- The frequency of an electromagnetic wave is determined by the oscillation frequency of the electric charges at the source of the wave.
- A variable capacitor and an inductor in a circuit provide one way to select the frequency of the desired electromagnetic wave, by adjusting the capacitor until the resonance frequency is reached.

24.2 THE ELECTROMAGNETIC SPECTRUM

- An electromagnetic wave, like any periodic wave, has a frequency f and a wavelength λ that are related to the speed υ of the wave by $\upsilon = f\lambda$.
- The ordered series of electromagnetic wave frequencies or wavelengths is called the **electromagnetic spectrum**.
- Although the boundary between frequency or wavelength regions is often shown as sharp, it

is not so well defined in practice, and the regions often overlap.

24.3 THE SPEED OF LIGHT

- The speed of light in a vacuum is

    $c = 299\ 792\ 458$ m/s

- Electromagnetic waves propagate through a vacuum at a speed given by

$$c = \frac{1}{\sqrt{\varepsilon_0 \mu_o}} \qquad (24.1)$$

where $\varepsilon_o = 8.85 \times 10^{-12}$ C^2/(N·m^2) is the (electric) permittivity of free space and $\mu_o = 4\pi \times 10^{-7}$ T·m/A is the (magnetic) permeability of free space.

24.4 THE ENERGY CARRIED BY ELECTROMAGNETIC WAVES

- Electromagnetic waves carry energy.

- Electric energy density $= \dfrac{1}{2}\varepsilon_o E^2$ (19.12)

- Magnetic energy density $= \dfrac{1}{2\mu_o} B^2$ (22.11)

- The **total energy density** u can be expressed in the following equivalent ways:

$$u = \frac{1}{2}\varepsilon_o E^2 + \frac{1}{2\mu_o} B^2 \qquad (24.2a)$$

$$u = \varepsilon_o E^2 \qquad (24.2b)$$

$$u = \frac{1}{\mu_o} B^2 \qquad (24.2c)$$

- The relation between the electric and magnetic fields in an electromagnetic wave is

$$E = cB \qquad (24.3)$$

- As an electromagnetic wave moves through space, it carries energy from one region to another. This energy transport is characterized by the **intensity** of the wave.
- It can be seen that the intensity S and the total energy density u are related as follows:

$$S = cu \qquad (24.4)$$

- The intensity of an electromagnetic wave depends on the electric and magnetic fields according to the following equivalent relations:

$$S = cu = \frac{1}{2} c \varepsilon_o E^2 + \frac{c}{2\mu_o} B^2 \qquad (24.5a)$$

$$S = c \varepsilon_o E^2 \qquad (24.5b)$$

$$S = \frac{c}{\mu_o} B^2 \qquad (24.5c)$$

24.5 THE DOPPLER EFFECT AND ELECTROMAGNETIC WAVES

- Electromagnetic waves also exhibit a Doppler effect, but it is different from that of sound waves. In the Doppler effect for electromagnetic waves, motion relative to a medium plays no role at all. Also, the speed at which electromagnetic waves travel has the same value, whether it is measured relative to a stationary observer or relative to one moving at a constant speed.
- When electromagnetic waves and the source and the observer of the waves all travel along the same line in a vacuum, the single equation that specifies the Doppler effect is

$$f' = f\left(1 \pm \frac{u}{c}\right) \text{ if } u \ll c \qquad (24.6)$$

In this expression, f' is the observed frequency, while f is the emitted frequency. The symbol u stands for the speed of the source and the observer relative to one another, and c is the speed of light in a vacuum. Equation 24.6 applies only if $u \ll c$. The plus sign in this equation applies when the source and the observer are moving toward one another, while the minus sign applies when they are moving apart.

24.6 POLARIZATION

- An electromagnetic wave is **linearly polarized**, when its vibrations always occur along one direction.
- The direction of polarization is taken arbitrarily to be that along which the electric field oscillates.
- Linearly polarized light can be produced from unpolarized light with the aid of certain materials. One commonly available material goes under the name of **Polaroid**. Such materials allow only the components of the electric field along one direction to pass through, while absorbing the field component perpendicular to this direction. The direction of polarization that a polarizing material allows through is called the **transmission axis**.
- No matter how the transmission axis is oriented, the intensity of the transmitted polarized light is one-half that of the incident unpolarized light. The reason for this is that

- the unpolarized light contains all polarization directions to an equal extent.
- Once polarized light has been produced with a piece of polarizing material, it is possible to use a second piece to change the polarization direction and to adjust the intensity of the light. The first piece of polarizing material is called the **polarizer**, while the second piece is referred to as the **analyzer**. The transmission angle of the analyzer is oriented at an angle θ relative to the transmission axis of the polarizer.
- The average intensity $\overline{S}$ of the light leaving the analyzer, then, is

$$\overline{S} = \overline{S}_o \cos^2 \theta \qquad (24.7)$$

where $\overline{S}_o$ is the average intensity of the light entering the analyzer. Equation 24.7 is sometimes called **Malus' law**.
- An application of a crossed polarizer/analyzer combination occurs in one kind of **liquid crystal display** (LCD).
- Polarized sunlight also originates from the scattering of light by molecules in the atmosphere.

CHAPTER 25 | ***THE REFLECTION OF LIGHT: MIRRORS***

The chapter discusses the reflection of light by mirrors.

The mirrors include spherical, as well as, plane mirrors, and the spherical mirrors may have the curvature in a concave or a convex manner. The treatment of curved mirrors requires the introduction of the concept of a focal point.

Important Concepts

- wave fronts
- rays
- plane waves
- law of reflection
- specular reflection
- diffuse reflection
- virtual image
- real image
- concave mirror
- convex mirror
- principal axis
- focal point

- focal length
- paraxial ray
- spherical aberration
- ray tracing
- principle of reversibility
- mirror equation

25.1 WAVE FRONTS AND RAYS

- Surfaces drawn through all points of a wave at the same phase of motion are called **wave fronts**.
- The radial lines pointing outward from the source and perpendicular to the wave fronts are called **rays**. The rays point in the direction of the velocity of the wave.
- Waves whose wave fronts are flat surfaces are known as **plane waves**.
- The concept of wave fronts can be applied to light waves.

25.2 THE REFLECTION OF LIGHT

- Most objects reflect a certain portion of the light falling on them.
- The **angle of incidence** θ_i is the angle that the incident ray makes with respect to the normal, which is a line drawn perpendicular to the surface at the point of incidence.
- The **angle of reflection** θ_r is the angle that the reflected ray makes with the normal.
- **The Law of Reflection**
  The incident ray, the reflected ray, and the normal to the surface all lie in the same plane, and the angle of reflection θ_r equals the angle of incidence θ_i:

  $$\theta_r = \theta_i$$

- When parallel light rays strike a smooth surface the reflected rays are parallel to each other. This type of reflection is known as **specular reflection**.
- The reflection from an irregular surface is known as **diffuse reflection**.

25.3 THE FORMATION OF IMAGES BY A PLANE MIRROR

- When you look into a plane mirror, you see an image of yourself that has three properties:
    1. The image is upright.
    2. The image is the same size as you are.
    3. The image is located as far behind the mirror as you are in front of it.
- Although rays of light appear to come from the image, it is evident that they do not originate from behind the plane mirror where the image appears to be. Because all of the rays of light do not actually emanate from the image, it is called a **virtual image**.
- Curved mirrors can produce images from which all the light rays do actually emanate. Such images are known as **real images**.

25.4 SPHERICAL MIRRORS

- If the inside or concave surface of the mirror is polished, it is a **concave mirror**.

- If the outside or convex surface is polished, it is a **convex mirror**.
- For either type of spherical mirror, the normal is drawn perpendicular to the mirror at the point of incidence. For each type, the center of curvature is labeled C, and the radius of curvature is labeled R. The **principal axis** of the mirror is a straight line drawn through C and the midpoint of the mirror.
- If an object is infinitely far away from a mirror, the rays from it are parallel to each other and to the principal axis as they approach the mirror. As these rays reflect, they pass through an image point that is referred to as the **focal point** F of the mirror.
- The distance between the focal point and the middle of the mirror is the **focal length** f of the mirror.
- For a concave mirror the focal length is one-half of the radius R:

$$f = \frac{1}{2}R \qquad (25.1)$$

- Rays that lie close to the principal axis are known as **paraxial rays**, and Equation 25.1 is valid only for such rays.

- The fact that a spherical mirror does not bring all rays parallel to the axis to a single image point is known as **spherical aberration**. Spherical aberration can be minimized by using a mirror whose height is small compared to the radius of curvature.
- A convex mirror also has a focal point.
- The focal point of a convex mirror is:

$$f = -\frac{1}{2}R \qquad (25.2)$$

25.5 THE FORMATION OF IMAGES BY SPHERICAL MIRRORS

- We can analyze the image produced by either concave or convex mirrors by using a graphical method called **ray tracing**.
- The **principle of reversibility** states that if the direction of a light ray is reversed, the light retraces its original path.
- A concave mirror can form a real or a virtual image, depending on where the object is located with respect to the mirror.

- A convex mirror always forms a virtual image of the object, no matter where in front of the mirror the object is placed.

25.6 THE MIRROR EQUATION AND THE MAGNIFICATION EQUATION

- The symbols used are:
  f = the focal length of the mirror
  d_o = the object distance, which is the distance between the object and the mirror
  d_i = the image distance, which is the distance between the image and the mirror
  m = the magnification of the mirror, which is the ratio of the height of the image to the height of the object.
- The **mirror equation** is:

$$\frac{1}{d_o} + \frac{1}{d_i} = \frac{1}{f} \qquad (25.3)$$

- The **magnification equation** is:

$$m = -\frac{d_i}{d_o} \qquad (25.4)$$

- The mirror and magnification equations can also be used with convex mirrors, provided the focal length f is taken to be a *negative number*.
- **Summary of Sign Conventions for Spherical Mirrors**

  Focal length

  f is + for concave mirror.

  f is − for convex mirror.

  Object distance

  d_o is + if the object is in front of the mirror (real image).

  d_o is − if the object is behind the mirror (virtual image).

  Image distance

  d_i is + if the image is in front of the mirror.

  d_i is − is the image is behind the mirror.

  Magnification

  m is + for an image that is upright with respect to the object.

  m is − for an image that is inverted with respect to the object.

CHAPTER 26 | ***THE REFRACTION OF LIGHT: LENSES AND OPTICAL INSTRUMENTS***

This chapter involves the refraction of light, which is quantified by Snell's law.

The concept of refraction is extended to a treatment of lenses, including optical devices such as microscopes and telescopes.

Important Concepts

- refraction
- Snell's law of refraction
- index of refraction
- critical angle
- total internal reflection
- Brewster angle
- dispersion
- focal point
- focal length
- converging lens
- diverging lens

- near point
- far point
- refractive power
- diopter
- angular magnification
- spherical aberration
- circle of least confusion
- chromatic aberration

26.1 THE INDEX OF REFRACTION

- The change of speed as a ray of light goes from one material to another causes the ray to deviate from its incident direction. This change in direction is called **refraction**
- To describe the extent to which the speed of light in a material medium differs from that in a vacuum, we use a parameter called the **index of refraction**.
- The index of refraction n is the ratio of the speed c of light in a vacuum to the speed v in a material medium:

$$n = \frac{c}{v} \qquad (26.1)$$

26.2 SNELL'S LAW AND THE REFRACTION OF LIGHT

- When light strikes the interface between two transparent materials the light generally divides into two parts. Part of the light is reflected, with an angle of reflection equal to the angle of incidence. The remainder of the light is transmitted across the interface.
- We label all variables associated with the incident (and reflected) ray with a subscript 1 and all variables associated with the refracted ray with a subscript 2.
- **Snell's Law of Refraction**
  When light travels from a material with refractive index n_1 into a material with refractive index n_2, the refracted ray, the incident ray, and the normal to the interface between the materials all lie in the same plane.

The angle of refraction θ_2 is related to the angle of incidence θ_1 by

$$n_1 \sin\theta_1 = n_2 \sin\theta_2 \qquad (26.2)$$

- The principle of conservation of energy indicates that the energy reflected plus the energy refracted must add up to equal the energy carried by the incident light, provided that none of the energy is absorbed by the material.
- The apparent depth when the observer is directly above a submerged object is

$$d' = d\left(\frac{n_2}{n_1}\right) \qquad (26.3)$$

- Although the incident and refracted waves have different speeds, they have the same frequency f.

26.3 TOTAL INTERNAL REFLECTION

- When light passes from a medium of larger refractive index to one of smaller refractive

index, the refracted ray bends *away* from the normal. When the angle of incidence reaches a certain value, called the **critical angle** θ_C, the angle of refraction is 90°. The refracted ray points along the surface.
- When the angle of incidence exceeds the critical angle, all the light is refracted back into the medium from which it came, a phenomenon called **total internal reflection**.
- The critical angle is defined by:

$$\sin\theta_C = \frac{n_2}{n_1} \quad (n_1 > n_2) \qquad (26.4)$$

- An important application of total internal reflection occurs in fiber optics, where hair-thin threads of glass or plastic, called optical fibers, "pipe" light from one place to another. The optical fiber consists of an inner core that carries the light and an outer concentric shell, the cladding.

26.4 POLARIZATION AND THE REFLECTION AND REFRACTION OF LIGHT

- There is one special angle of incidence at which the reflected light is completely polarized parallel to the surface, the refracted ray being only partially polarized. This angle is called the **Brewster angle** θ_B.

$$\tan \theta_B = \frac{n_2}{n_1} \quad (26.5)$$

This equation is called **Brewster's law**.

26.5 THE DISPERSION OF LIGHT: PRISMS AND RAINBOWS

- The refractive index depends on the wavelength, and the rays corresponding to different colors are bent by different amounts

by a prism and depart traveling in different directions.
- The spreading of light into its color components is called **dispersion**.

26.6 LENSES

- Rays that are near the principal axis and parallel to it converge to a single point on the axis after emerging from the lens. This point is called the **focal point** F of the lens.
- The distance between the focal point and the lens is the **focal length** f.
- In what follows we assume the lens is so thin compared to f that it makes no difference whether f is measured between the focal point and either surface of the lens or the center of the lens.
- One type of lens is known as a **converging lens**, because it causes incident parallel rays to converge at the focal point.
- A **diverging lens** causes incident rays to diverge after exiting the lens.

26.7 THE FORMATION OF IMAGES BY LENSES

- Each point on an object emits light rays in all directions, and when some of these rays pass through a lens, they form an image.
- We can analyze the image produced by a converging lens by using a graphical method called ray tracing.
- The image formed by a converging lens may be real or virtual, depending on where the object is located with respect to the lens.
- Light rays diverge upon leaving a diverging lens. Regardless of the position of a real object, a diverging lens always forms a virtual image that is upright and smaller relative to the object.

26.8 THE THIN-LENS EQUATION AND THE MAGNIFICATION EQUATION

- The thin-lens equation is

$$\frac{1}{d_0} + \frac{1}{d_i} = \frac{1}{f} \qquad (26.6)$$

- The magnification equation is

$$m = \frac{h_i}{h_o} = -\frac{d_i}{d_o} \qquad (26.7)$$

- Summary of Sign Conventions for Lenses
  Focal length
  f is + for converging lens.
  f is - for diverging lens.
  Object distance
  d_o is + if the object is to the left of the lens (real object), as is usual.
  d_o is - if the object is to the right of the lens (virtual object).
  Image distance
  d_i is + for an image (real) formed to the right of the lens by a real object.
  d_i is - for an image (virtual) formed to the left of the lens by a real object.
  Magnification
  m is + for an image that is upright with respect to the object.
  m is - for an image that is inverted with respect to the object.

26.9 LENSES IN COMBINATION

- Many optical instruments, such as microscopes and telescopes, use a number of lenses together to produce an image. Among other things, a multiple-lens system can produce an image that is magnified more than is possible with a single lens.
- The image produced by one lens serves as the object for the next lens.

26.10 THE HUMAN EYE

- Light enters the eye through a transparent membrane (the cornea). This membrane covers a clear liquid region (the aqueous humor), behind which is a diaphragm (the iris), the lens, a region filled with a jelly-like substance (the vitreous humor), and, finally, the retina. The retina is the light-sensitive part of the eye, consisting of millions of structures called rods and cones. When stimulated by light, these structures send electrical impulses via the optic nerve to the brain, which interprets the image on the retina. The variable opening at the center of the iris is called the pupil.

- For clear vision, the eye must refract the incoming light rays, so as to form a sharp image on the retina. In reaching the retina, the light passes through five different media, each with a different index of refraction.
- **Accommodation** is the process in which the lens of the eye changes its focal point.
- The point nearest the eye at which an object can be placed and still produce a sharp image on the retina is called the **near point** of the eye.
- The **far point** of the eye is the location of the farthest object on which the fully relaxed eye can focus.
- A person who is nearsighted (myopic) can focus on nearby objects but cannot clearly see objects far away.
- A farsighted (hyperopic) person can usually see distant objects clearly, but cannot focus on those nearby.
- The **refractive power** describes the extent to which a lens refracts light.

Refractive power of a lens = $1/f$ (f in meters)
$$(26.8)$$

- The refractive power is measured in units of **diopters**. One diopter is 1 m^{-1}.

26.11 ANGULAR MAGNIFICATION AND THE MAGNIFYING GLASS

- The angle θ that is subtended by the object at the eye is called the **angular size** of both the image and the object.
- The **angular magnification** M is the angular size θ' of the final image produced by the instrument divided by a reference angular size θ.

$$M = \frac{\theta'}{\theta} \qquad (26.9)$$

- For a magnifying glass, the angular magnification is

$$M = \frac{\theta'}{\theta} \approx \left(\frac{1}{f} - \frac{1}{d_i}\right)N \qquad (26.10)$$

where N is the distance between the near point and the eye.

26.12 THE COMPOUND MICROSCOPE

- To increase the angular magnification beyond that possible with a magnifying glass, an additional converging lens can be included to "premagnify" the object before the magnifying glass comes into play. The result is an optical instrument known as the **compound microscope**. The magnifying glass is called the eyepiece, and the additional lens is called the objective. The angular magnification of a compound microscope is

$$M \approx -\frac{(L - f_e)N}{f_o f_e} \quad (26.11)$$

In Equation 26.11, f_o and f_e are the focal lengths of the objective and the eyepiece, respectively, and L is the distance between the lenses.

26.13 THE TELESCOPE

- A telescope is an instrument for magnifying distant objects.
- The angular magnification of an astronomical telescope is

$$M = \frac{\theta'}{\theta} \approx -\frac{f_o}{f_e} \qquad (26.12)$$

- The **refracting telescope** includes a lens for the eyepiece.
- The **reflecting telescope** uses a concave mirror for the objective.

26.14 LENS ABERRATIONS

- One common type of aberration is **spherical aberration**, and it occurs with converging and diverging lenses made with spherical surfaces.
- The **circle of least confusion** is where the most satisfactory image can be formed by a lens.

- **Chromatic aberration** arises because the index of refraction of the material from which the lens is made varies in wavelength.
- Chromatic aberration can be greatly reduced using a compound lens.

CHAPTER 27 | ***INTERFERENCE AND THE WAVE NATURE OF LIGHT***

This chapter considers the combination of several waves of light, and the constructive and destructive interference that this produces.

The concept of coherent light is presented, because interference is best observed with coherent waves.

The interference of light is essential to understanding diffraction patterns and the instrument known as an interferometer.

Important Concepts

- constructive interference
- destructive interference
- coherent sources
- interferometer
- diffraction
- Huygens' principle
- resolving power
- diffraction grating
- principal fringe

27.1 THE PRINCIPLE OF LINEAR SUPERPOSITION

- When two or more light waves pass through a given point, their electric fields combine according to the principle of linear superposition to give the resultant electric field at that point.
- Interference can and does alter the brightness of light.
- When two identical waves (same wavelength and same amplitude) arrive at a point in phase, that is crest to crest and trough to trough, the waves reinforce each other and **constructive interference** occurs.
- When two identical waves arrive at a point out of phase with one another, or crest to trough, they mutually cancel and **destructive interference occurs**.
 If constructive or destructive interference is to continue at a point, the sources must be **coherent sources**. Two sources are coherent sources it the waves they emit maintain a constant phase relation.

27.2 YOUNG'S DOUBLE-SLIT EXPERIMENT

- For a double slit, the angle θ for the interference maxima (bright fringes) can be determined from the following expression:

$$\sin\theta = m\frac{\lambda}{d} \quad m = 0, 1, 2, 3,... \quad (27.1)$$

The angle for the interference for the minima (dark fringes) can be determined from the following expression:

$$\sin\theta = \left(m+\frac{1}{2}\right)\frac{\lambda}{d} \quad m = 0, 1, 2, 3,... \quad (27.2)$$

27.3 THIN-FILM INTERFERENCE

- When monochromatic light strikes a film nearly perpendicularly, both reflection and refraction occur. Part of the light reflects from the front surface of the film and part from the back surface. Part of the light reflected from the back surface exits the front surface, combining with the light reflected from that surface. Because of the extra travel distance and the possibility of phase shifts encountered at the two surfaces, there can be constructive or destructive interference between the two waves.

- The wavelength that is important for thin-film interference is the wavelength within the film, not the wavelength in a vacuum.

$$\lambda_{film} = \frac{\lambda_{vacuum}}{n} \qquad (27.3)$$

where n is the index of refraction of the film.

27.4 THE MICHELSON INTERFEROMETER

- An **interferometer** is an apparatus that can be used to measure the wavelength of light by utilizing interference between two light waves.
- Waves emitted by a monochromatic light source strike a beam splitter, so called because it splits the beam into two parts. One wave strikes an adjustable mirror and reflects back on itself. The other wave strikes a fixed mirror and returns. The two reflected waves combine

and reach the telescope. The viewer sees constructive or destructive interference, depending only on the difference in path lengths traveled by the two waves.

27.5 DIFFRACTION

- **Diffraction** is the bending of waves around obstacles or the edges of an opening.
- **Huygens' principle** states that:
  Every point on a wave front acts as a source of tiny wavelets that move forward with the same speed as the wave; the wave front at a later instant is the surface that is tangent to the wavelets.
- Huygens' principle applies to all kinds of waves.
- The extent to which a wave bends around the edges of an opening is determined by the ratio λ/W, where λ is the wavelength of the wave and W is the width of the opening.
- The dark fringes for single-slit diffraction are at an angle θ according to

$$\sin\theta = m\frac{\lambda}{W} \quad m = 1, 2, 3,\ldots \qquad (27.4)$$

- Between each pair of dark fringes there is a bright fringe due to constructive interference.
- In the production of computer chips it is important to minimize the effects of diffraction. Miniaturization is achieved using the techniques of photolithography.

27.6 RESOLVING POWER

- The **resolving power** of an optical instrument is its ability to distinguish between two closely spaced objects.
- Diffraction patterns place a natural limit on the resolving power of an instrument.
- The diffraction pattern created by a small circular opening when the viewing screen is far from the opening consists of a bright central region, surrounded by alternating dark and bright circular fringes. The angle θ locates the first circular dark fringe relative to the central bright region and is given by

$$\sin\theta = 1.22\frac{\lambda}{D} \qquad (27.5)$$

where λ is the wavelength of the light and D is the diameter of the opening.
- Two point objects are just resolved when the first dark fringe in the diffraction pattern of one falls directly on the central bright fringe in the diffraction pattern of the other.
- The Rayleigh criterion for resolution is

$$\theta_{min} \approx 1.22 \frac{\lambda}{D} \quad (\theta_{min} \text{ in radians}) \qquad (27.6)$$

For a given wavelength λ and aperture diameter D, this result specifies the smallest angle that two point objects can subtend at the aperture and still be resolved.

27.7 THE DIFFRACTION GRATING

- Fringe patterns also result when light passes through more than two slits, and the arrangement consisting of a large number of parallel, closely spaced slits is called a **diffraction grating**.
- Each bright fringe is located by an angle θ relative to the central fringe. These bright fringes are sometimes called **principal fringes** or **principal maxima**.

- Constructive interference creates the principal fringes.
- The principal maxima of a diffraction grating are at:

$$\sin\theta = m\frac{\lambda}{d} \quad m = 0, 1, 2, 3,\ldots \quad (27.7)$$

- A diffraction grating produces bright fringes that are much narrower or sharper than those from a double slit.

27.8 COMPACT DISCS AND THE USE OF INTERFERENCE

- The audio information on a CD is encoded in the form of raised areas on the bottom of the disc. These raised areas appear as "pits" when viewed from the top. They are separated by flat areas called "land." The pits and land are covered by a transparent coating.

27.9 X-RAY DIFFRACTION

- A diffraction pattern results when X-rays are directed onto a crystalline material. The pattern consists of a complicated arrangement of spots, because the crystal has a complex three-dimensional structure.

CHAPTER 28 | *SPECIAL RELATIVITY*

This chapter views the concepts of mechanics when an object travels at a speed comparable to the speed of light. The relativistic shifts of time, length, momentum, and energy are considered.

Important Concepts

- event
- coordinate system
- special relativity
- inertial reference frame
- relativity postulate
- speed of light postulate
- time dilation
- proper time interval
- proper length
- relativistic momentum
- total energy
- pair production
- velocity-addition formula

28.1 EVENTS AND INERTIAL REFERENCE FRAMES

- In the theory of relativity, an **event** is a physical "happening" that occurs at a certain place and time.
- To record the event, each observer uses a **reference frame** that consists of a set of x, y, z axes (called a **coordinate system**) and a clock.
- The theory of **special relativity** deals with a "special" kind of reference frame, called an **inertial reference frame**. An inertial reference frame is one in which Newton's law of inertia is valid.
- Rotating and otherwise accelerating reference frames are not inertial reference frames.

28.2 THE POSTULATES OF SPECIAL RELATIVITY

- **The Postulates of Special Relativity**
  1. **The Relativity Postulate**. The laws of physics are the same in every inertial

reference frame.
2. **The Speed of Light Postulate.** The speed of light in a vacuum, measured in any inertial reference frame, always has the same value of c, no matter how fast the source of light and the observer are moving relative to each other.

- Any inertial reference frame is as good as any other for expressing the laws of physics.
- There is no experiment that can distinguish between an inertial frame that is at rest and one that is moving at a constant velocity.
- It is not possible to single out one particular reference frame as being at "absolute rest."
- When an object moves slowly (v is much smaller than c), the modification of classical concepts is negligibly small, and the classical view of each concept provides an accurate description of reality. However, when the object moves so rapidly that v is an appreciable fraction of the speed of light, the effects of the speed of light must be considered.

28.3 THE RELATIVITY OF TIME: TIME DILATION

- **Time dilation** is given by

$$\Delta t = \frac{\Delta t_o}{\sqrt{1 - \frac{v^2}{c^2}}} \qquad (28.1)$$

  The symbols in this formula are summarized as follows:
  Δt_o = proper time interval between two events, as measured by an observer who is at rest with respect to the events and who views them as occurring at the same place
  Δt = time interval measured by an observer who is in motion with respect to the events and who views them as occurring at different places
  v = relative speed between two observers
  c = speed of light in a vacuum
- Being at rest with respect to a clock is the usual or "proper" situation, so the time interval Δt_o is called the **proper time interval**.

28.4 THE RELATIVITY OF LENGTH: LENGTH CONTRACTION

- The length contraction is given by

$$L = L_o \sqrt{1 - \frac{v^2}{c^2}} \qquad (28.2)$$

The length L_o is called the **proper length**; it is the length (or distance) between two points as measured by an observer at rest with respect to them.

28.5 RELATIVISTIC MOMENTUM

- The theory of special relativity reveals that the magnitude of the **relativistic momentum** must be as defined in Equation 28.3:

$$p = \frac{mv}{\sqrt{1 - \frac{v^2}{c^2}}} \qquad (28.3)$$

28.6 THE EQUIVALENCE OF MASS AND ENERGY

- The **total energy** E of a moving object is related to its mass and speed by the following relation:

$$E = \frac{mc^2}{\sqrt{1 - \frac{v^2}{c^2}}} \quad (28.4)$$

- When $v = 0$, the total energy is called the **rest energy** E_o, and Equation 28.4 reduces to

$$E_o = mc^2 \quad (28.5)$$

- We can write the kinetic energy as

$$KE = E - E_o = mc^2 \left(\frac{1}{\sqrt{1 - \frac{v^2}{c^2}}} - 1 \right) \quad (28.6)$$

- Since mass and energy are equivalent, any change in one is accompanied by a corresponding change in the other.
- The transformation of electromagnetic waves into matter happens. The process in which a gamma ray is transformed into two particles is known as **pair production**.
- One of the important consequences of the theory of special relativity is that objects with mass cannot reach the speed of light c in a vacuum. Thus, the speed of light represents the ultimate speed.

28.7 THE RELATIVISTIC ADDITION OF VELOCITIES

- The theory of relativity states that the velocity u of an object measured in one inertial reference frame is related to the velocity u' measured in another inertial reference frame by the **velocity-addition formula**:

$$u = \frac{u'+v}{1+\dfrac{u'v}{c^2}} \tag{28.7}$$

In this equation the symbols have the following meaning:

u = the velocity of the object as measured relative to an inertial reference frame

u' = the velocity of the object measured relative to a second inertial reference frame that is moving at a velocity v relative to the first one

CHAPTER 29 | *PARTICLES AND WAVES*

This chapter examines the dual nature of both matter and electromagnetic waves. The introduction of the photon and the use of Planck's constant to quantify the relationship between electromagnetic waves and matter are of particular importance.

The major result of the Heisenberg uncertainty principle is eventually addressed, incorporating the concept of momentum into the chapter.

Important Concepts

- wave-particle duality
- Planck's constant
- quantization
- blackbody radiation
- photon
- photoelectric effect
- photoelectron
- work function
- photodiode
- photoevaporation
- Compton effect

- Compton wavelength of the electron
- de Broglie wavelength
- wave function
- quantum mechanics
- Heisenberg uncertainty principle

29.1 THE WAVE-PARTICLE DUALITY

- Particles can behave like waves and exhibit interference effects.
- Scientists now accept the **wave-particle duality** as an essential part of nature: Waves can exhibit particle-like characteristics, and particles can exhibit wave-like characteristics.

29.2 BLACKBODY RADIATION AND PLANCK'S CONSTANT

- All bodies, no matter how hot or cold, continually radiate electromagnetic waves.
- Planck assumed that the energy E of an atomic oscillator could have only discrete values:

$$E = nhf \quad n = 0, 1, 2, 3,... \quad (29.1)$$

where n is zero or a positive integer, f is the frequency of vibrations (in hertz), and h is a constant now called **Planck's constant**.

- Planck's constant is $h = 6.626\,0755 \times 10^{-34}$ J·s.
- Whenever the energy of a system can have only certain definite values, and nothing in between, the energy is said to be **quantized**.
- Conservation of energy requires that the energy carried off by the radiated electromagnetic waves must equal the energy lost by the atomic oscillators in Planck's model.
- Planck's model for **blackbody radiation** sets the stage for the idea that electromagnetic energy occurs as a collection of discrete packets of energy, the energy of a packet being equal to hf.

29.3 PHOTONS AND THE PHOTOELECTRIC EFFECT

- Experimental evidence that light consists of **photons** comes from a phenomenon called the **photoelectric effect**, in which electrons are emitted from a metal surface when light shines

on it. The electrons are emitted if the light has a sufficiently high frequency.
- Because the electrons are ejected with the aid of light, they are called **photoelectrons**.
- Light of frequency f can be regarded as a collection of discrete packets of energy (photons), each packet containing an amount of energy E given by

$$E = hf \qquad (29.2)$$

where h is Planck's constant.
- If the photon has enough energy to do the work of removing an electron from a metal, the electron can be ejected. For the least strongly held electrons, the necessary work has a minimum value W_o and is called the **work function** of the metal. If the photon has energy in excess of the work needed to remove an electron, the excess energy appears as kinetic energy of the ejected electron. The following equation describes the photoelectric effect:

$$hf = KE_{max} + W_o \qquad (29.3)$$

- Only light with a frequency above a certain minimum value $f_o = W_o/h$ will eject electrons. If the frequency of the light is below this value,

no electrons are ejected, regardless of how intense the light is.
- Another significant feature of the photoelectric effect is that the maximum kinetic energy of the ejected electrons remains the same when the intensity of the light increases, provided the light frequency remains the same.
- A photon travels at the speed of light in a vacuum, and does not exist as an object at rest.
- The energy of a photon is entirely kinetic in nature, for it has no rest energy and no mass.
- A **photodiode** is a type of *pn* junction diode.
- **Photoevaporation** is the process in which high-energy, ultraviolet (UV) photons from hot stars outside a cloud heat it up.

29.4 THE MOMENTUM OF A PHOTON AND THE COMPTON EFFECT

- The process in which an X-ray photon is scattered from an electron, the scattered photon having a smaller frequency than the incident photon, is called the **Compton effect**. The Compton effect is described by the relation

$$hf = hf' + KE \qquad (29.4)$$

where f' is the frequency of the scattered photon.

- For an initially stationary electron, conservation of linear momentum requires that

Momentum of incident photon =
Momentum of scattered photon +
Momentum of recoil electron   (29.5)

- The momentum of a photon is

$$p = \frac{hf}{c} = \frac{h}{\lambda} \qquad (29.6)$$

where λ is the wavelength of the photon.

- The difference between the wavelength λ' of the scattered photon and the wavelength λ of the incident photon is related to the scattering angle θ by

$$\lambda' - \lambda = \frac{h}{mc}(1 - \cos\theta) \qquad (29.7)$$

In this equation m is the mass of the electron. The quantity h/mc is referred to as the **Compton wavelength of the electron**, and has the value $h/mc = 2.43 \times 10^{-12}$ m.

- The photoelectric effect and the Compton effect provide compelling evidence that light can exhibit particle-like characteristics attributable to energy packets called photons.

29.5 THE DE BROGLIE WAVELENGTH AND THE WAVE NATURE OF MATTER

- De Broglie made the explicit proposal that the wavelength λ of a particle is given by the same relation (Equation 29.6) that applies to a photon:

$$\lambda = \frac{h}{p} \quad (29.8)$$

where h is Planck's constant and p is the magnitude of the relativistic momentum of the particle. Today, λ is known as the **de Broglie wavelength** of the particle.
- Particles other than electrons can also exhibit wave-like properties.
- Particle waves are waves of probability, waves whose magnitude at a point in space gives an indication of the probability that the particle will be found at that point.
- Ψ is referred to as the **wave function** of the particle.

29.6 THE HEISENBERG UNCERTAINTY PRINCIPLE

- The uncertainty relation

$$\Delta p_y \approx \frac{h}{W} \qquad (29.9)$$

indicates that a smaller slit width W leads to a larger uncertainty Δp_y in the y component of an electron's momentum.

- **The Heisenberg Uncertainty Principle**
  Momentum and position

$$(\Delta p_y)(\Delta y) \geq \frac{h}{2\pi} \qquad (29.10)$$

Δy = uncertainty in a particle's position along the *y* direction
Δp_y = uncertainty in the *y* component of the linear momentum of the particle
Energy and time

$$(\Delta E)(\Delta t) \geq \frac{h}{2\pi} \qquad (29.11)$$

ΔE = uncertainty in the energy of the particle when the particle is in a certain state
Δt = time interval during which the particle is in the state

The Heisenberg uncertainty principle states that it is impossible to specify precisely both the momentum and the position of a particle at the same time.

CHAPTER 30 | ***THE NATURE OF THE ATOM***

This chapter discusses various models of the atom that have been considered, and the details of how each fits or does not fit current theory. The electrons that are associated with atoms, and their energies, are of particular interest. Much attention is given to a quantum approach to orbital electrons.

A main concept regarding electrons is the Pauli exclusion principle.

The emission of electromagnetic waves as electrons change to lower energy states is also a major topic. Consequences of such emissions, particularly the laser, are presented in detail.

Important Concepts

- nucleus
- line spectra
- Balmer series
- Lyman series
- Paschen series
- Rydberg constant
- stationary orbits
- Bohr model
- Bohr radius
- energy levels

- ground state
- excited state
- ionization energy
- emission lines
- absorption lines
- principal quantum number
- orbital quantum number
- magnetic quantum number
- spin quantum number
- probability cloud
- shell
- subshell
- Pauli exclusion principle
- periodic table
- X-ray
- Bremsstrahlung
- characteristic lines
- laser
- spontaneous emission
- stimulated emission
- population inversion
- metastable
- holography
- hologram

30.1 RUTHERFORD SCATTERING AND THE NUCLEAR ATOM

- An atom contains a small, positively charged **nucleus**, which is surrounded at relatively large distances by a number of electrons.
- In the natural state, an atom is electrically neutral, because the nucleus contains a number of protons (each with a charge $+e$) that equals the number of electrons (each with a charge $-e$).
- Rutherford concluded that the positive charge, instead of being distributed thinly and uniformly throughout the atom, was concentrated in a small region called the nucleus.

30.2 LINE SPECTRA

- The continuous range of emitted wavelengths is characteristic of a solid. In contrast, individual atoms, free of the strong interactions that are

present in a solid, emit only specific wavelengths, rather than a continuous range.
- The individual wavelengths emitted by a gas can be separated and identified as a series of bright fringes or lines. The series of lines is called a **line spectrum**.
- The group of lines known as the **Balmer series** is found from the empirical equation that gives the values of the observed wavelengths. This equation is given below, along with similar equations that apply to the **Lyman series** and the **Paschen series**.

Lyman series $\quad \dfrac{1}{\lambda} = R\left(\dfrac{1}{1^2} - \dfrac{1}{n^2}\right) \quad n = 2, 3, \ldots$

(30.1)

Balmer series $\dfrac{1}{\lambda} = R\left(\dfrac{1}{2^2} - \dfrac{1}{n^2}\right) \quad n = 3, 4, 5, \ldots$

(30.2)

Paschen series $\dfrac{1}{\lambda} = R\left(\dfrac{1}{3^2} - \dfrac{1}{n^2}\right) \quad n = 4, 5, 6, \ldots$

(30.3)

In these equations, the constant term R has the value of $R = 1.097 \times 10^7 \text{ m}^{-1}$ and is called the **Rydberg constant**.

30.3 THE BOHR MODEL OF THE HYDROGEN ATOM

- The electron orbits are called **stationary orbits** or stationary states.
- Bohr recognized that radiationless orbits violated the laws of physics, as they were then known. But the assumption of such orbits was necessary, because the traditional laws indicated that an electron radiates electromagnetic waves as it accelerates around a circular path, and the loss of the energy carried by the waves would lead to the collapse of the orbit.
- Bohr recognized that a photon is emitted only when the electron *changes* orbits from a larger one with a higher energy E_i to a smaller one with a lower energy E_f.

$$E_i - E_f = hf \qquad (30.4)$$

- We assume that the nucleus contains Z electrons, so the total energy E of the hydrogen atom is

$$E = \text{KE} + \text{EPE} = \frac{1}{2}mv^2 - \frac{kZe^2}{r} \qquad (30.5)$$

- But a centripetal force must act. Therefore,

$$mv^2 = \frac{kZe^2}{r} \qquad (30.6)$$

- We can use this result to eliminate the term mv^2 from Equation 30.5, with the result that

$$E = \frac{1}{2}\left(\frac{kZe^2}{r}\right) - \frac{kZe^2}{r} = -\frac{kZe^2}{2r} \qquad (30.7)$$

- The total energy of the atom is negative, because the negative electric potential energy is larger in magnitude than the positive kinetic energy.
- Bohr conjectured that the angular momentum L can assume only discrete values; in other words, L is quantized. He postulated that the allowed values are integer multiples of Planck's constant divided by 2π:

$$L_n = mv_n r_n = n\,\frac{h}{2\pi} \quad n = 1, 2, 3,\ldots \qquad (30.8)$$

- The radius r_n of the nth Bohr orbit is given by

$$r_n = \left(\frac{h^2}{4\pi^2 mke^2}\right)\frac{n^2}{Z} \quad n = 1, 2, 3,\ldots \quad (30.9)$$

- The radii of the Bohr orbits (in meters) are:

$$r_n = (5.29 \times 10^{-11})\frac{n^2}{Z} \quad n = 1, 2, 3,\ldots \quad (30.10)$$

- The smallest Bohr orbit ($n = 1$) has a radius of $r_1 = 5.29 \times 10^{-11}$ m. This particular value is called the **Bohr radius**.
- The total energy for the nth orbit is

$$E_n = -\left(\frac{2\pi^2 mk^2 e^4}{h^2}\right)\frac{Z^2}{n^2} \quad n = 1, 2, 3,\ldots (30.11)$$

- The Bohr energy levels in joules are

$$E_n = -(2.18 \times 10^{-18} \text{ J})\frac{Z^2}{n^2} \quad n = 1, 2, 3,\ldots (30.12)$$

- Often, atomic energies are expressed in electron volts rather than joules. The Bohr energy levels in electron volts are

$$E = -(13.6 \text{ eV})\frac{Z^2}{n^2} \quad n = 1, 2, 3,... \quad (30.13)$$

- It is useful to represent the energies given by Equation 30.13 on an **energy level diagram**. The lowest energy level is called the **ground state**, to distinguish it from the higher levels, which are called the **excited states**.
- Energies of the excited states come closer and closer together as n increases.
- The energy needed to remove an electron from the ground ($n = 1$) state and place it infinitely far from the atom is called the **ionization energy**.
- To predict the spectra of the hydrogen atom, Bohr combined his result for the energy with Einstein's idea of the photon. The wavelength of the emitted photon is

$$\frac{1}{\lambda} = \left(\frac{2\pi^2 mk^2 e^4}{h^3 c}\right)(Z^2)\left(\frac{1}{n_f^2} - \frac{1}{n_i^2}\right) \quad (30.14)$$

$n_i, n_f = 1, 2, 3,...$ and $n_i > n_f$

- The various lines in the hydrogen atom spectrum are produced when electrons change from higher to lower energy levels and photons are emitted. Consequently, the spectral lines are called **emission lines**. Electrons can also make transitions in the reverse direction, from lower to higher levels, in a process known as **absorption**.

- If photons with a continuous range of wavelengths pass through a gas and then are analyzed with a grating spectroscope, a series of dark **absorption lines** appears in the continuous spectrum. The dark lines indicate the wavelengths that have been removed by the absorption process.
- The Bohr model is now known to be oversimplified and has been superseded by a more detailed picture provided by quantum mechanics and the Schrodinger equation.

30.4 DE BROGLIE'S EXPLANATION OF BOHR'S ASSUMPTION ABOUT ANGULAR MOMENTUM

- In de Broglie's way of thinking, the electron in its circular Bohr orbit must be pictured as a particle wave. And like waves traveling on a string, particle waves can lead to standing waves under resonant conditions.
- The total distance around a Bohr orbit of radius r is the circumference of the orbit, or $2\pi r$. By the same reasoning, then, the condition for

standing particle waves for the electron in a Bohr orbit would be

$$2\pi r = n\lambda \quad n = 1, 2, 3,...$$

where n is the number of whole wavelengths that fit into the circumference of the circle.

30.5 THE QUANTUM MECHANICAL PICTURE OF THE HYDROGEN ATOM

- Quantum mechanics reveals that four different quantum numbers are needed to describe each state of the hydrogen atom. These four are described below:
    1. **The principal quantum number n.** As in the Bohr model, this number determines the total energy of the atom and can have only integer values.
    2. **The orbital quantum number ℓ.** This number determines the angular momentum L of the electron due to its orbital motion.

    $$L = \sqrt{\ell(\ell+1)}\,\frac{h}{2\pi} \quad (30.15)$$

$\ell = 0, 1, 2, ..., (n-1)$

3. **The magnetic quantum number m_ℓ.** The word "magnetic" is used here because an externally applied magnetic field influences the energy of the atom, and this quantum number is used in describing the effect. This effect is known as the Zeeman effect. The magnetic quantum number determines the component of the angular momentum along a specific direction, which is called the z direction by convention.

4. **The spin quantum number m_s.** This number is needed because the electron has an intrinsic property called spin angular momentum. Loosely speaking, we can view the electron as spinning while it orbits the nucleus. There are two possible values for the spin quantum number of the electron:

$m_s = +1/2$ or $m_s = -1/2$

Sometimes the phrases "spin up" and "spin down" are used to refer to the directions of the spin angular momentum associated with the values for m_s.

- If we make a large number of measurements of the electron's position with respect to the nucleus, we would find an electron's position is uncertain, in the sense that there is a probability of finding the electron at a distance from the nucleus. The probability is determined by the wave function Ψ.
- A picture is constructed from so many measurements that the individual dots are no longer visible, but have merged to form a kind of probability "**cloud**" whose density changes gradually from place to place. The dense regions indicate places where the probability of finding an electron is higher, while the less dense regions indicate places where the probability is lower.
- The cloud is spherical for an electron in the $n = 1$ state.
- The cloud for $n = 2$ and $\ell = 1$ has a two-lobe shape with the nucleus at the center between the lobes. For larger values of n, the probability clouds become increasingly complex and are spread out over larger volumes of space.

30.6 THE PAULI EXCLUSION PRINCIPLE AND THE PERIODIC TABLE OF THE ELEMENTS

- Except for hydrogen, all electrically neutral atoms contain more than one electron, the number being given by the atomic number Z of the element. In addition to being attracted by the nucleus, the electrons repel each other. This repulsion contributes to the total energy of a multiple-electron atom.
- The simplest approach to dealing with a multiple-electron atom still uses the four quantum numbers.
- Detailed quantum mechanical calculations reveal that the energy level of each state of a multiple-electron atom depends on both the principle quantum number n and the orbital quantum number ℓ. The energy generally increases as n increases. For a given n, the energy also increases as ℓ increases, but there are some exceptions.
- In a multiple-electron atom, all electrons with the same value of n are said to be in the same **shell**.
- Those electrons with the same values for both n and ℓ are said to be in the same **subshell**.

- The lowest energy state for an atom is called the **ground state**.
- **The Pauli Exclusion Principle**
  No two electrons in an atom can have the same set of values for the four quantum numbers n, ℓ, m_ℓ, and m_s.
- Because of the Pauli exclusion principle, there is a maximum number of electrons that can fit into each subshell.
- Each entry in the **periodic table** of the elements often includes the ground state electronic configuration.
- Group 0, the last column of elements on the right side of the table, consists of the noble gases.
- Group I is made up of the alkali metals.
- Group VII consists of the halogens.
- By looking at the other groups (II-VI) of the periodic table, you can see that elements within a group have outermost electrons in either s or p subshells.

30.7 X-RAYS

- **X-rays** can be produced when electrons, accelerated through a large potential difference, collide with a metal target made, for example, from molybdenum or platinum.
- The broad continuous spectrum is referred to as **Bremsstrahlung** and is emitted when the electrons decelerate or "brake" upon hitting the target.
- The sharp peaks in the X-ray spectrum are called the **characteristic X-rays**.
- The X-ray spectrum has a sharp cutoff that occurs at a wavelength of λ_o on the short wavelength side of the Bremsstrahlung. This wavelength is

$$\lambda_o = \frac{hc}{eV} \qquad (30.17)$$

where V is the potential difference through which the electron is accelerated.

30.8 THE LASER

- When an electron makes a transition from a higher energy state to a lower energy state, a photon is emitted.
- In **spontaneous emission**, a photon is emitted spontaneously, in a random direction, without external provocation.
- In **stimulated emission**, an incoming photon induces or stimulates the electron to change energy levels. To produce stimulated emission, however, the incoming photon must have an energy that exactly matches the difference between the energies of the two levels, namely, $E_i - E_f$.
- Stimulated emission has three important features. First, one photon goes in and two photons come out. Second, the emitted photon travels in the same direction as the incoming photon. And third, the emitted photon is exactly in step with, or has the same phase as, the incoming photon. In other words, the two electromagnetic waves that these two photons represent are coherent.
- While stimulated emission plays a pivotal role in a **laser**, other factors are also important. For instance, an external source of energy must be

available to excite electrons into higher energy levels.
- If sufficient energy is delivered to the atoms, more electrons will be excited to a higher energy level than remain in a lower energy level, a condition known as a **population inversion**. The population inversions used in lasers involve a higher energy state that is **metastable**, in the sense that electrons remain in it for a much longer period of time than they do in an ordinary excited state. The requirement of a metastable higher energy state is essential, so that there is more time to enhance the population inversion.
- When the stimulated emission of a laser involves only a single pair of energy levels, the output beam has a single frequency or wavelength and is said to be monochromatic.
- A laser beam is narrow.

30.9 MEDICAL APPLICATIONS OF THE LASER

- One of the medical areas in which the laser has had a substantial impact is in ophthalmology,

which deals with the structure, function, and diseases of the eye.
- Another medical application of the laser is in the treatment of congenital capillary malformations or port-wine stains.
- In the treatment of cancer, the laser is being used along with light-activated drugs in photodynamic therapy.

30.10 HOLOGRAPHY

- On of the most familiar applications of lasers is in **holography**, which is a process for producing three-dimensional images. The information used to produce a holographic image is captured on a photographic film, which is referred to as a **hologram**.
- A holographic image differs greatly from a photographic image. The most obvious difference is that a hologram provides a three-dimensional image, while photographs are two dimensional. The reason that holographic images are three dimensional is inherent in the interference pattern formed on the film.

CHAPTER 31 | ***NUCLEAR PHYSICS AND RADIOACTIVITY***

This chapter is concerned with the nucleus, and its stability. The composition of the nucleus, rather than the electrons, is the central topic.

The forces in the nucleus, and the radioactive decay that occurs when these forces are unable to hold the nucleus together, are also considered.

Important Concepts

- nucleus
- nucleon
- neutron
- proton
- atomic number
- atomic mass number (nucleon number)
- isotope
- strong nuclear force
- radioactivity
- binding energy
- mass defect
- atomic mass unit
- α, β, and γ rays
- radioactive decay

- α decay
- parent nucleus
- daughter nucleus
- transmutation
- β decay
- positron
- γ decay
- neutrino
- weak nuclear force
- electroweak force
- half-life
- activity
- decay constant
- radioactive decay series
- Geiger counter
- scintillation counter
- semiconductor detector
- cloud chamber
- bubble chamber
- photographic emulsion

31.1 NUCLEAR STRUCTURE

- Atoms consist of electrons in orbit about a central **nucleus**.

- The nucleus of an atom consists of **neutrons** and **protons**, collectively referred to as **nucleons**. The neutron carries no electrical charge and has a mass slightly larger than that of a proton.
- The number of protons in the nucleus is different in different elements and is given by the **atomic number** Z. In an electrically neutral atom, the number of nuclear protons equals the number of electrons in orbit around the nucleus. The total number of protons and neutrons is referred to as the **atomic mass number** A.

$$A = Z + N \qquad (31.1)$$

Sometimes A is called the **nucleon number**.
- Nuclei that contain the same number of protons, but different numbers of neutrons, are known as **isotopes**.
- The protons and neutrons in the nucleus are clustered together to form an approximately spherical region. The radius r of the nucleus depends on the atomic mass number A and is given approximately in meters by

$$r \approx (1.2 \times 10^{-15} \text{ m}) A^{1/3} \qquad (31.2)$$

31.2 THE STRONG NUCLEAR FORCE AND THE STABILITY OF THE NUCLEUS

- Two positive charges that are as close together as they are in a nucleus repel one another with a very strong electrostatic force.
- The gravitational force of attraction between nucleons is too weak to counteract the repulsive electric force, so it must be that a different type of force holds the nucleus together. This force is the **strong nuclear force** and is one of only three fundamental forces that have been discovered. The gravitational force is also one of these forces, as is the electroweak force (see Section 31.5).
- Many features of the strong nuclear force are known:
    It is almost independent of electric charge.
    The range of action of the strong nuclear force is extremely short.
- The limited range of action of the strong nuclear force plays an important role in the stability of the nucleus. For a nucleus to be stable, the electrostatic repulsion between the protons must be balanced by the attraction

- between the nucleons due to the strong nuclear force.
- As the atomic number Z of protons in the nucleus increases, the number N of neutrons has to increase even more, if stability is to be maintained.
- As more and more protons occur in a nucleus, there comes a point when a balance of repulsive and attractive forces cannot be achieved by an increased number of neutrons.
- All nuclei with more than 83 protons are unstable and spontaneously break apart or rearrange their internal structure as time passes. This spontaneous disintegration or rearrangement of internal structure is called **radioactivity**.

31.3 THE MASS DEFECT OF THE NUCLEUS AND NUCLEAR BINDING ENERGY

- The more stable the nucleus is, the more energy is needed to break it apart. The required energy is called the **binding energy** of the nucleus.

- Two ideas come into play as we discuss the binding energy of the nucleus. These are mass and the rest energy of an object.
- The sum of the individual masses of the separated protons and neutrons is greater by an amount Δm than the mass of the intact nucleus. The difference in mass Δm is known as the **mass defect** of the nucleus. The binding energy of the nucleus is related to the mass defect via

$$\text{Binding energy} = (\text{Mass defect})c^2 = (\Delta m)c^2$$
(31.3)

- In calculations it is customary to use the **atomic mass unit** (u) instead of the kilogram.
- Figure 31.6 shows a graph in which the binding energy divided by the nucleon number A is plotted against the nucleon number itself. The peak for the ^4_2He isotope of helium indicates that the ^4_2He nucleus is particularly stable. The binding energy per nucleon increases rapidly for nuclei with small masses and reaches a maximum of approximately 8.7 MeV/nucleon for a nucleon number of about $A = 60$. For greater nucleon numbers, the binding energy per nucleon decreases gradually. Eventually, the binding energy per nucleon decreases

enough so there is insufficient binding energy to hold the nucleus together.

31.4 RADIOACTIVITY

- When an unstable or radioactive nucleus disintegrates spontaneously, particles and/or high energy photons, collectively called "**rays**," are released.
- Three kinds of rays are produced by naturally occurring radioactivity: α rays, β rays, and γ rays. They are named according to their ability to penetrate matter:
  α rays are the least penetrating.
  γ rays are the most penetrating.
- In all **radioactive decay** processes it has been observed that the number of nucleons present before the decay is equal to the number of nucleons present after the decay.
- When a nucleus disintegrates and produces α rays, it is said to undergo α **decay**.
- α rays consist of positively charged particles, each one being the $^{4}_{2}$He nucleus of helium. An α particle has a charge of $+2e$ and a nucleon number of $A = 4$.

- The original nucleus is referred to as the **parent nucleus**, and the nucleus remaining after disintegration is called the **daughter nucleus**.
- α decay converts one element into another, a process known as **transmutation**.
- When α decay occurs, the energy released appears as kinetic energy of the recoiling nucleus and the α particle, except for a small portion carried away as a γ ray.
- β^- particles are electrons.
- β^- decay causes a transmutation of one element into another.
- The electron emitted in β^- decay does not actually exist within the parent nucleus and is not one of the orbital electrons. Instead, the electron is created when a neutron decays into a proton and an electron. The electron is usually fast-moving and escapes from the atom.
- A second kind of β decay sometimes occurs. In this process the particle emitted by the nucleus is a **positron**. A positron, also called a β^+ particle, has the same mass as an electron, but carries a charge of $+e$ instead of $-e$.
- The emitted positron does not exist within the nucleus, but, rather, is created when a nuclear proton is transformed into a neutron.
- The nucleus, like the orbital electrons, exists only in discrete energy states or levels. When a

nucleus changes from an excited state to a lower energy state, a photon (γ ray) is emitted.
- γ decay does not cause a transmutation of one element into another.

31.5 THE NEUTRINO

- When a β particle is emitted by a radioactive nucleus, energy is simultaneously released. However, it is found that most β particles do not have enough kinetic energy to account for all the energy released.
- Part of the energy is carried away by another particle that is emitted along with the β particle. This additional particle is called the **neutrino**.
- The neutrino has zero electrical charge.
- The emission of neutrinos and β particles involves a force called the **weak nuclear force**, because it is much weaker than the strong nuclear force. The weak nuclear force and the electromagnetic force are two different manifestations of a single, more fundamental force, the **electroweak force**.

31.6 RADIOACTIVE DECAY AND ACTIVITY

- The **half-life** of a radioactive isotope is the time required for one-half of the nuclei present to disintegrate.
- The **activity** of a radioactive sample is the number of disintegrations per second, $\Delta N/\Delta t$, that occurs:

$$\frac{\Delta N}{\Delta t} = -\lambda\ N \qquad (31.4)$$

   where λ is a proportionality constant referred to as the **decay constant**. The minus sign is present in this equation because each disintegration decreases the number N of nuclei originally present.
- The SI unit for activity is the **becquerel** (Bq); one becquerel equals one disintegration per second. Activity is also measured in terms of a unit called the **curie** (Ci):

$$1\ \text{Ci} = 3.70 \times 10^{10}\ \text{Bq}$$

- The number N of radioactive nuclei present at time t is

$$N = N_o e^{-\lambda t} \qquad (31.5)$$

assuming that the number at $t = 0$ is N_o.

- The half-life $T_{1/2}$ is related to the decay constant λ by

$$T_{1/2} = \frac{\ln 2}{\lambda} = \frac{0.693}{\lambda} \qquad (31.6)$$

31.7 RADIOACTIVE DATING

- The radiocarbon dating technique utilizes the $^{14}_{6}C$ isotope of carbon.
- Other methods utilize uranium $^{238}_{92}U$, potassium $^{40}_{19}K$, and lead $^{210}_{82}Pb$. For such methods to be useful, the half-life of the radioactive species must be neither too short nor too long relative to the age of the sample to be dated.

31.8 RADIOACTIVE DECAY SERIES

- When an unstable nucleus decays, the resulting daughter nucleus is sometimes also unstable. If

so, the daughter then decays and produces its own daughter, and so on, until a completely stable nucleus is produced. This sequential decay of one nucleus after another is called a **radioactive decay series**.
- Ultimately, the series ends with lead $^{206}_{82}Pb$.

31.9 DETECTORS OF RADIATION

- The most familiar detector is the **Geiger counter**. A wire electrode runs along the center of the tube and is kept at a high positive voltage relative to the outer cylinder. When a high-energy particle or photon enters the cylinder, it collides with and ionizes a gas molecule. The number of counts or clicks is related to the number of high-energy particles or photons present, or equivalently, to the number of disintegrations that produced the particles or photons.
- The **scintillation counter** is another important radiation detector. Often the scintillator is a crystal containing a small amount of impurity, but plastic, liquid, and gaseous scintillators are also used. In response to ionizing radiation, the scintillator emits a flash of visible light. The

photons of the flash then strike the photocathode of the photomuliplier tube.

- Ionizing radiation can also be detected with several types of **semiconductor detectors**. Such devices utilize *n*- and *p*- type materials, and their operation depends on the electrons and holes formed in the materials as a result of the radiation. One of the main advantages of semiconductor detectors is their ability to discriminate between two particles with only slightly different energies.
- In a **cloud chamber**, a gas is cooled just to the point where it will condense into droplets, provided nucleating agents are available on which the droplets can form.
- A **bubble chamber** works in a similar fashion, except it contains a liquid that is just at the point of boiling. Tiny bubbles form along the trail of a high-energy particle passing through the liquid.
- A **photographic emulsion** also can be used directly to produce a record of the path taken by a particle of ionizing radiation.

CHAPTER 32 | *IONIZING RADIATION, NUCLEAR ENERGY, AND ELEMENTARY PARTICLES*

This chapter is concerned with the biological consequences of exposure to radiation, the use of radioactivity in reactors, and the extremely small particles, especially the quarks.

The idea that protons and neutrons are not the simplest forms of matter, and the charge of an electron is not the most basic charge unit, are among the concepts now accepted in understanding the nucleus.

Important Concepts

- ionizing radiation
- exposure
- roentgen
- absorbed dose
- gray
- rad
- relative biological effectiveness
- biologically equivalent dose
- rem

- radiation sickness
- induced nuclear transmutation
- transuranium elements
- thermal neutrons
- nuclear fission
- chain reaction
- uncontrolled fission chain reaction
- controlled fission chain reaction
- fuel element
- reactor core
- moderator
- subcritical
- supercritical
- control rod
- nuclear fusion
- elementary particles
- antiparticle
- photon family
- lepton family
- hadron family
- meson
- baryon
- quark
- charmed
- color
- the standard model
- cosmology
- Hubble's law
- Hubble parameter

- expanding universe
- Big Bang theory
- standard cosmological model
- GUT force
- Grand Unified Theory
- cosmic background radiation

32.1 BIOLOGICAL EFFECTS OF IONIZING RADIATION

- **Ionizing radiation** consists of photons and/or moving particles that have sufficient energy to knock an electron out of an atom or molecule, thus forming an ion. The photons usually lie in the ultraviolet, X-ray, or γ-ray regions of the electromagnetic spectrum, while the moving particles can be the α and β particles emitted during radioactive decay.
- Nuclear radiation is potentially harmful to humans, because the ionization it produces can alter significantly the structure of molecules within living cells. The alterations can lead to the death of the cell and even the organism itself. Despite the potential hazards, ionizing radiation is used in medicine for diagnostic and therapeutic purposes.

- **Exposure** is a measure of the ionization produced in air by X-rays or γ-rays.
- Exposure = q/m, where q is the total charge of the positive ions produced and m is the mass of dry air being ionized.
- The SI unit for exposure is the coulomb per kilogram (C/kg).
- With q expressed in coulombs (C) and m in kilograms (kg), the exposure in **roentgens** (R) is given by

$$\text{Exposure (in roentgens)} = \left(\frac{1}{2.58 \times 10^{-4}}\right)\frac{q}{m}$$

(32.1)

- For biological purposes, the **absorbed dose** is a more suitable quantity, because it is the energy absorbed from the radiation per unit mass of absorbing material:

$$\text{Absorbed dose} = \frac{\text{Energy Absorbed}}{\text{Mass of absorbing material}}$$

(32.2)

- The SI unit of absorbed dose is the **gray** (Gy), which is the unit of energy divided by the unit of mass: 1 Gy = 1 J/kg.

- The word **rad** is an acronym for radiation absorbed dose. The rad and the gray are related by
  $$1 \text{ rad} = 0.01 \text{ gray}$$
- To compare the damage caused by different types of radiation, the **relative biological effectiveness** (RBE) is used:

$$\text{Relative biological effectiveness} = \frac{\text{The dose of 200-keV X-rays that produces a certain biological effect}}{\text{The dose of radiation that produces the same biological effect}} \quad (32.3)$$

- The RBE depends on the nature of the ionizing radiation and its energy, as well as the type of tissue being irradiated.
- The product of the absorbed dose in rads (not in grays) and the RBE is the **biologically equivalent dose**:

$$\text{Biologically equivalent dose (in rem)} = \text{Absorbed dose (in rad)} \times \text{RBE} \quad (32.4)$$

- The unit for the biologically equivalent dose is the **rem**, short for roentgen equivalent, man.
- Everyone is continually exposed to background radiation from natural sources.

- **Radiation sickness** is the general term applied to the acute effects of radiation.
- Long term or latent effects of radiation may appear as a result of high-level brief exposure or low-level exposure over a long period of time.

32.2 INDUCED NUCLEAR REACTIONS

- A **nuclear reaction** is said to occur whenever the incident nucleus, particle, or photon causes a change to occur in a target nucleus.
- If an incident particle induces a transmutation, the reaction is called an **induced nuclear transmutation**.
- Induced transmutations can be used to produce isotopes that are not found naturally.
- Elements with a higher atomic number than uranium ($Z = 92$) are known as **transuranium elements**. None of the transuranium elements occurs naturally.
- The neutrons that participate in nuclear reactions that have a kinetic energy of about 0.04 eV or less are called **thermal neutrons**.

32.3 NUCLEAR FISSION

- The splitting of a massive nucleus into two less-massive fragments is known as **nuclear fission**.
- The energy carried off by the fragments is enormous and was stored in the original nucleus primarily in the form of electric potential energy.
- The fact that the uranium fission reaction releases 2.5 neutrons, on the average, makes it possible for a **self-sustaining series of fissions** to occur.
- A **chain reaction** is a series of nuclear fissions whereby some of the neutrons produced by each fission cause additional fissions.
- With an average energy of about 200 MeV being released per fission, an **uncontrolled** chain reaction can generate an incredible amount of energy in a very short time.
- By limiting the number of neutrons in the environment of the fissile nuclei, it is possible to establish a condition whereby each fission event contributes, on the average, only one neutron that fissions another nucleus. In this manner the chain reaction and the rate of

energy production are **controlled**. The controlled fission chain reaction is the principle behind nuclear reactors used in the commercial generation of power.

32.4 NUCLEAR REACTORS

- The **fuel elements** contain the fissile fuel.
- The entire region of fuel elements is known as the **reactor core**.
- The material that slows down the neutrons is called the **moderator**.
- The reactor is **subcritical** when, on average, the neutrons from each fission trigger less than one subsequent fission.
- When the neutrons from each fission trigger more than one additional fission, the reactor is **supercritical**.
- A control mechanism is needed to keep a reactor in its normal or critical state. This control is accomplished by a number of **control rods** that can be moved into and out of the reactor core. The control rods contain an element, such as boron or cadmium, that readily absorbs neutrons without fissioning.

32.5 NUCLEAR FUSION

- Two very-low-mass nuclei with relatively small binding energies per nucleon can be combined or "fused" into a single, more massive nucleus that has a greater binding energy per nucleon. This process is called **nuclear fusion**. A substantial amount of energy can be released during a fusion reaction.
- Reactions that require extremely high temperatures are called **thermonuclear reactions**.

32.6 ELEMENTARY PARTICLES

- Several hundred nuclear particles exist, and scientists no longer believe that the proton and the neutron are elementary particles.
- For every type of particle there is a corresponding type of antiparticle. The **antiparticle** is a form of matter that has the same mass as the particle, but carries an

opposite electric charge or a magnetic moment that is oriented in an opposite direction relative to the spin. A few electrically neutral particles, like the photon and the neutral pion, are their own antiparticles.

- The **photon family** has only one member, the photon.
- The **lepton family** consists of particles that interact by means of the *weak nuclear force*.
- The **hadron family** contains the particles that interact by means of the *strong nuclear force and the weak nuclear force*.
- The hadrons are subdivided into two groups, the **mesons** and the **baryons**.
- Hadrons are made up of smaller, more elementary particles called **quarks**.
- Three quarks are named *up* (u), *down* (d), and *strange* (s) and are assigned to have, respectively, fractional charges of $(+2/3)e$, $-(1/3)e$, and $-(1/3)e$.
- A quark possesses a charge magnitude smaller than that of an electron.
- Three more quarks are named *charmed* (c), *top* (t), and *bottom* (b).
- Each quark is believed to posses a characteristic called **color**, for which there are three possibilities: blue, green, or red. The corresponding possibilities for antiquarks are antiblue, antigreen, and antired. The use of the

term "color" and the specific choices of blue, green, and red is arbitrary.
- The quark property called color is an important one, for it brings the quark model into agreement with the Pauli exclusion principle.
- In particle physics the phrase "**the standard model**" refers to the currently accepted explanation for the strong nuclear force, the weak nuclear force, and the electromagnetic force.

32.7 COSMOLOGY

- **Cosmology** is the study of the structure and evolution of the universe.
- Hubble found that a galaxy located at a distance d from the earth recedes from the earth at a speed v given by

  Hubble's law $\qquad v = Hd \qquad (32.5)$

  where H is a constant known as the **Hubble parameter**.
- Hubble's picture of an **expanding universe** does not mean that the earth is at the center of the expansion.

- The idea that our galaxy and other galaxies in the universe were very close together at some earlier instant in time lies at the heart of the **Big Bang theory**. This theory postulates that the universe had a definite beginning in a cataclysmic event.
- Scientists have proposed an evolutionary sequence of events following the Big Bang. This sequence of events is known as the **standard cosmological model.**
- Immediately after the Big Bang, there was only one force, the unified force. In time, however, this force produced the four forces that we see today (the electromagnetic force, the weak nuclear force, the strong nuclear force, and the gravitational force), as indicated in Figure 32.16.
- As the temperature of the universe decreased further, stars and galaxies formed, and today we find a temperature of 2.7 K characterizing the **cosmic background radiation** of the universe.

NOTES

NOTES

NOTES

NOTES

NOTES

NOTES

NOTES

NOTES